Ben Stacy Jerrik (Ed.)

Sprengelia

Ben Stacy Jerrik (Ed.)

Sprengelia

Ericaceae, Shrub, Family, Heather, Plant, Botany

Part Press

Imprint

Permission is granted to copy, distribute and/or modify this document under the terms of the GNU Free Documentation License, Version 1.2 or any later version published by the Free Software Foundation; with no Invariant Sections, with the Front-Cover Texts, and with the Back- Cover Texts. A copy of the license is included in the section entitled "GNU Free Documentation License".

All parts of this book are extracted from Wikipedia, the free encyclopedia (www.wikipedia.org).

You can get detailed informations about the authors of this collection of articles at the end of this book. The editors (Ed.) of this book are no authors. They have not modified or extended the original texts.

Pictures published in this book can be under different licences than the GNU Free Documentation License. You can get detailed informations about the authors and licences of pictures at the end of this book.

The content of this book was generated collaboratively by volunteers. Please be advised that nothing found here has necessarily been reviewed by people with the expertise required to provide you with complete, accurate or reliable information. Some information in this book maybe misleading or wrong. The Publisher does not guarantee the validity of the information found here. If you need specific advice (f.e. in fields of medical, legal, financial, or risk management questions) please contact a professional who is licensed or knowledgeable in that area.

Any brand names and product names mentioned in this book are subject to trademark, brand or patent protection and are trademarks or registered trademarks of their respective holders. The use of brand names, product names, common names, trade names, product descriptions etc. even without a particular marking in this works is in no way to be construed to mean that such names may be regarded as unrestricted in respect of trademark and brand protection legislation and could thus be used by anyone.

Cover image: www.ingimage.com
Concerning the licence of the cover image please contact ingimage.

Publisher:
Part Press is a trademark of
International Book Market Service Ltd., 17 Rue Meldrum, Beau Bassin, 1713-01 Mauritius
Email: info@bookmarketservice.com
Website: www.bookmarketservice.com

Published in 2012

Printed in: U.S.A., U.K., Germany. This book was not produced in Mauritius.

ISBN: 978-613-8-88918-2

Contents

Sprengelia

<table>
<tr><td colspan="2" align="center">***Sprengelia***</td></tr>
<tr><td colspan="2" align="center">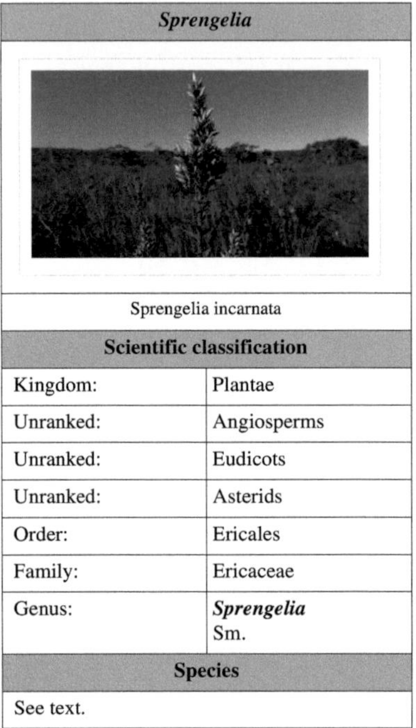</td></tr>
<tr><td colspan="2" align="center">Sprengelia incarnata</td></tr>
<tr><td colspan="2" align="center">**Scientific classification**</td></tr>
<tr><td>Kingdom:</td><td>Plantae</td></tr>
<tr><td>Unranked:</td><td>Angiosperms</td></tr>
<tr><td>Unranked:</td><td>Eudicots</td></tr>
<tr><td>Unranked:</td><td>Asterids</td></tr>
<tr><td>Order:</td><td>Ericales</td></tr>
<tr><td>Family:</td><td>Ericaceae</td></tr>
<tr><td>Genus:</td><td>***Sprengelia***
Sm.</td></tr>
<tr><td colspan="2" align="center">**Species**</td></tr>
<tr><td colspan="2">See text.</td></tr>
</table>

Sprengelia is a genus of shrubs in the family Ericaceae.[1]

The four species are all endemic to Australia:

- *Sprengelia distichophylla* (Rodway) W.M.Curtis
- *Sprengelia incarnata* Sm. - Pink Swamp-heath
- *Sprengelia monticola* (DC.) Druce - Rock Sprengelia
- *Sprengelia sprengelioides* (R.Br.) Druce

The genus was first formally described by English botanist James Edward Smith in 1794 in *Konglia Vetenskaps Academiens Nya Handlingar*.[2]

References

[1] *"Sprengelia"* (http://plantnet.rbgsyd.nsw.gov.au/cgi-bin/NSWfl.pl?page=nswfl&lvl=gn&name=Sprengelia). *PlantNET - New South Wales Flora Online*. Royal Botanic Gardens & Domain Trust, Sydney Australia. . Retrieved 2010-06-14.

[2] *"Sprengelia"* (http://www.anbg.gov.au/cgi-bin/apni?TAXON_NAME=Sprengelia). *Australian Plant Name Index (APNI), IBIS database*. Centre for Plant Biodiversity Research, Australian Government, Canberra. . Retrieved 2010-06-14.

Ericaceae

<table>
<tr><th colspan="2" style="text-align:center">Ericaceae</th></tr>
<tr><td colspan="2">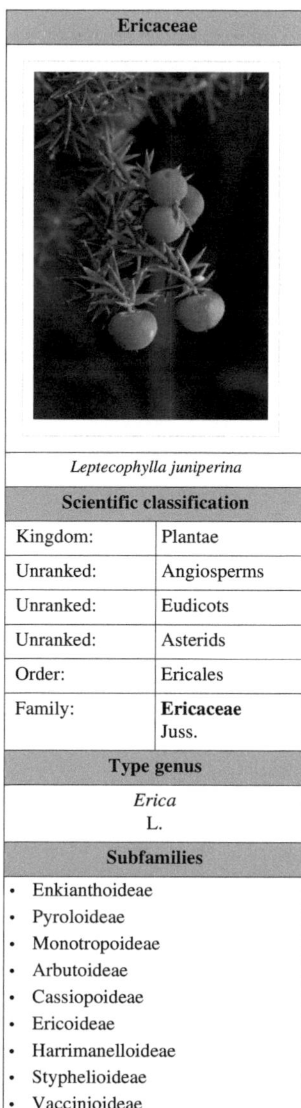</td></tr>
<tr><td colspan="2" style="text-align:center">Leptecophylla juniperina</td></tr>
<tr><th colspan="2" style="text-align:center">Scientific classification</th></tr>
<tr><td>Kingdom:</td><td>Plantae</td></tr>
<tr><td>Unranked:</td><td>Angiosperms</td></tr>
<tr><td>Unranked:</td><td>Eudicots</td></tr>
<tr><td>Unranked:</td><td>Asterids</td></tr>
<tr><td>Order:</td><td>Ericales</td></tr>
<tr><td>Family:</td><td>Ericaceae
Juss.</td></tr>
<tr><th colspan="2" style="text-align:center">Type genus</th></tr>
<tr><td colspan="2" style="text-align:center">Erica
L.</td></tr>
<tr><th colspan="2" style="text-align:center">Subfamilies</th></tr>
<tr><td colspan="2">

- Enkianthoideae
- Pyroloideae
- Monotropoideae
- Arbutoideae
- Cassiopoideae
- Ericoideae
- Harrimanelloideae
- Styphelioideae
- Vaccinioideae

</td></tr>
</table>

The **Ericaceae**, commonly known as the **heath** or **heather** family, is a group of mostly calcifuge (lime-hating) flowering plants. The family is large, with roughly 4000 species spread across 126 genera, making it the 14th most speciose family of flowering plants.[1] There are many well-known and economically important members of the Ericaceae, these include the cranberry, blueberry, huckleberry, azalea, rhododendron, and various common heaths and heathers (*Erica*, *Cassiope*, *Daboecia*, and *Calluna* for example).[2]

Description

The Ericaceae contains a morphologically diverse range of taxa, including herbs, dwarf shrubs, shrubs and trees. The leaves are usually alternate or whorled, simple and without stipules, and hermaphrodite flowers. The flowers show considerable variability. The petals are often fused (sympetalous) with shapes ranging from narrowly tubular to funnelform or widely bowl-shaped. The corollas are usually radially symmetrical (actinomorphic) but many flowers of the genus *Rhododendron* are somewhat bilaterally symmetrical (zygomorphic).[3]

Distribution and ecology

Ericads have a nearly worldwide distribution. They are absent from continental Antarctica, parts of the high Arctic, central Greenland, northern and central Australia, and much of the lowland tropics and neotropics.[1]

The family is largely made up of calcifuge plants, that tend to thrive only in acidic soils. This is a trait not found in the Clethraceae and Cyrillaceae, the two families most closely related to the Ericaceae.

Most Ericaceae (excluding the Monotropoideae, Pyroloideae, and some Styphelioideae) form mycorrhizae, where fungi grow in and around the roots and provide the plant with nutrients. This symbiotic relationship is considered crucial to the success of members of the family in edaphically stressful environments worldwide.[4] The Pyroleae tribe are mixotrophic and gain sugars from the mycorrhizae as well as nutrients.[5]

In many parts of the world, a "heath" or "heathland" is an environment characterised by an open dwarf-shrub community found on low quality acidic soils, generally dominated by plants in the Ericaceae. In eastern North America, members of this family often grow in association with an oak canopy, in a type of ecology known as an oak-heath forest.[6] [7]

Systematics

In 2002 systematic research conducted by Kron *et al.*[8] resulted in the inclusion of the formerly recognised families Empetraceae, Epacridaceae, Monotropaceae, Prionotaceae and Pyrolaceae into the Ericaceae. This was based on a combination of molecular, morphological, anatomical, and embryological data, analysed within a phylogenetic framework. The move significantly increased the morphological and geographical range found within the group. The resulting family now includes 9 subfamilies, 126 genera, and c. 4000 species:

1. Enkianthoideae Kron, Judd & Anderberg (1 genus, 16 species)
2. Pyroloideae Kosteltsky (4 genera, 40 species)
3. Monotropoideae Arnott (10 genera, 15 species)
4. Arbutoideae Niedenzu (5 genera, 80 species)
5. Cassiopoideae Kron & Judd (1 genus, 12 species)
6. Ericoideae Link (19 genera, 1790 species)
7. Harrimanelloideae Kron & Judd (1 genus, 2 species)
8. Styphelioideae Sweet (35 genera, 545 species)
9. Vaccinioideae Arnott (50 genera, 1580 species)

Etymology

The name Ericaceae comes from the type genus *Erica*, which appears derived from the Greek word *ereike*. The exact meaning is difficult to interpret, but some sources show it as simply meaning 'heather.'[9] The name may have been used informally to refer to the plants in pre-Linnaean times, and was simply formalised when Linnaeus described *Erica* in 1753, and then when Jussieu described the Ericaceae in 1789.[10]

Genera

See the full list at *List of Ericaceae genera.*

References

[1] Stevens, P. F. (2001 onwards). Angiosperm Phylogeny Website. Version 9, June 2008. http://www.mobot.org/MOBOT/research/APweb/

[2] Kathleen A. Kron, E. Ann Powell and J. L. Luteyn (2002). "Phylogenetic relationships within the blueberry tribe (Vaccinieae, Ericaceae) based on sequence data from MATK and nuclear ribosomal ITS regions, with comments on the placement of Satyria" (http://www.amjbot. org/cgi/content/full/89/2/327). *American Journal of Botany* **89** (2): 327–336. doi:10.3732/ajb.89.2.327. PMID 21669741. .

[3] Watson, L., Dallwitz, M.J. (1992 onwards) The families of flowering plants: descriptions, illustrations, identification, and information retrieval. Version: 4th March 2011. http://delta-intkey.com.

[4] Cairney, JWG & Meharg, AA (2003). "Ericoid mycorrhiza: a partnership that exploits harsh edaphic conditions". *European Journal of Soil Science* **54** (4): 735–740. doi:10.1046/j.1351-0754.2003.0555.x.

[5] Error: Bad DOI specified!

[6] *The Natural Communities of Virginia Classification of Ecological Community Groups* (Version 2.3), Virginia Department of Conservation and Recreation, 2010 (http://www.dcr.virginia.gov/natural_heritage/ncTIIIe.shtml)

[7] Schafale, M. P. and A. S. Weakley. 1990. *Classification of the natural communities of North Carolina: third approximation.* North Carolina Natural Heritage Program, North Carolina Division of Parks and Recreation.

[8] Kron, K.A., Judd, W.S., Stevens, P.F., Crayn, D.M., Anderberg, A.A., Gadek, P.A., Quinn, C.J., Luteyn, J.L. (2002). "Phylogenetic Classification of Ericaceae: Molecular and Morphological Evidence". *The Botanical Review* **68** (3): 335–423.

[9] Wiktionary. 2011. Ericaceae. http://en.wiktionary.org/wiki/Ericaceae

[10] Jussieu, A.-L. de. 1789. Genera plantarum ordines naturales disposita. pg. 159-160. Herissant & Barrois, Paris.

External links

- Ericaceae (http://www.theplantlist.org/browse/A/Ericaceae/) at *The Plant List* (http://www.theplantlist. org)
- Ericaceae (http://delta-intkey.com/angio/www/ericacea.htm), Epacridaceae (http://delta-intkey.com/angio/ www/epacrida.htm), Empetraceae (http://delta-intkey.com/angio/www/empetrac.htm), Monotropaceae (http://delta-intkey.com/angio/www/monotrop.htm), and Pyrolaceae (http://delta-intkey.com/angio/www/ pyrolace.htm) at *The Families of Flowering Plants (DELTA)* (http://delta-intkey.com/angio/)
- Ericaceae (http://eol.org/pages/4269/overview) at the *Encyclopedia of Life* (http://eol.org/)
- Ericaceae (http://www.mobot.org/mobot/research/APweb/orders/ericalesweb.htm#Ericaceae) at the *Angiosperm Phylogeny Website* (http://www.mobot.org/mobot/research/APweb/)
- Ericaceae (http://www.efloras.org/florataxon.aspx?flora_id=1&taxon_id=10316) at the online *Flora of North America* (http://www.efloras.org/flora_page.aspx?flora_id=1)
- Ericaceae (http://www.efloras.org/florataxon.aspx?flora_id=2&taxon_id=10316) at the online *Flora of China* (http://www.efloras.org/flora_page.aspx?flora_id=2)
- Ericaceae (http://www.efloras.org/florataxon.aspx?flora_id=5&taxon_id=10316) at the online *Flora of Pakistan* (http://www.efloras.org/flora_page.aspx?flora_id=5)
- Ericaceae (http://www.efloras.org/florataxon.aspx?flora_id=60&taxon_id=10316) at the online *Flora of Chile* (http://www.efloras.org/flora_page.aspx?flora_id=60)
- Epacridaceae (http://floraseries.landcareresearch.co.nz/pages/Taxon. aspx?id=_16c46185-ad3b-4db4-86b3-fd283d59f582&fileName=Flora 1.xml) at the online *Flora of New Zealand* (http://floraseries.landcareresearch.co.nz/pages/index.aspx)

- Epacridaceae (http://florabase.dec.wa.gov.au/browse/profile.php/22885) at the online *Flora of Western Australia* (http://florabase.dec.wa.gov.au/)
- Ericaceae (http://www.ericaceae.org/homepage.html) at *Ericaceae.org* (http://www.ericaceae.org/)
- Ericaceae (http://keys.trin.org.au:8080/key-server/data/0e0f0504-0103-430d-8004-060d07080d04/media/Html/taxon/Ericaceae.htm) at the *Australian Tropical Rainforest Plants Identification System* (http://www.anbg.gov.au/cpbr/cd-keys/rfk/)
- Neotropical Blueberries (http://www.nybg.org/bsci/res/lut2) at the *New York Botanical Garden* (http://www.nybg.org/)

Ericales

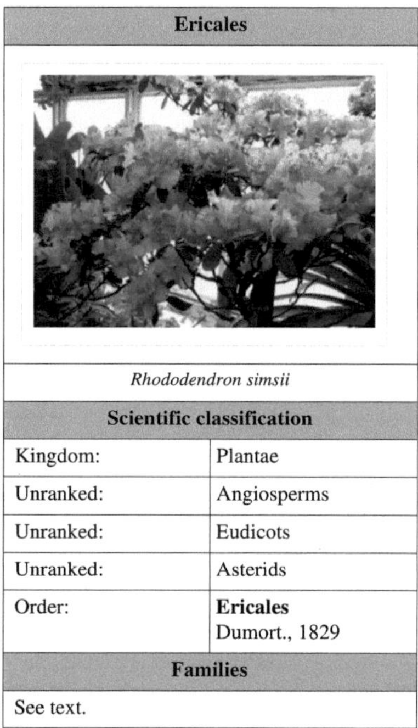

	Ericales
Rhododendron simsii	
Scientific classification	
Kingdom:	Plantae
Unranked:	Angiosperms
Unranked:	Eudicots
Unranked:	Asterids
Order:	**Ericales** Dumort., 1829
Families	
See text.	

The **Ericales** are a large and diverse order of dicotyledons, including for example tea, persimmon, blueberry, Brazil nut, and azalea. The order includes trees and bushes, lianas and herbaceous plants. Together with ordinary autophytic plants, the Ericales include chlorophyll-deficient myco-heterotrophic plants (e. g. *Sarcodes sanguinea*) and carnivorous plants (e. g. genus *Sarracenia*).

Many species have five petals, often grown together. Fusion of the petals is a trait that was traditionally used to place the order in the subclass Sympetalae.[1]

Mycorrhiza is an interesting property, frequently associated with the Ericales. Indeed, the symbiosis with root fungi is quite common among the order representatives, and there are even three kinds of it which can be found exclusively among Ericales (namely, ericoid, arbutoid and monotropoid mycorrhiza). In addition, some families among the order are notable for their exceptional ability to accumulate aluminum (Jansen et al., 2004).

Ericales are a cosmopolitic order. Areas of distribution of families vary largely - while some are restricted to tropics, others exist mainly in Arctic or temperate regions. The entire order contains over 8000 species, of which the Ericaceae account for 2000-4000 species (by various estimates).

Economic importance

The most profitable plant in the order is tea (*Camellia sinensis*) from the Theaceae family. The order also includes some edible fruits, including kiwi fruit (*Actinidia deliciosa*), persimmon (genus *Diospyros*), blueberry, huckleberry and cranberry, Brazil nut, and Mamey sapote. The order also includes shea (*Vitellaria paradoxa*), which is the major dietary lipid source for millions of sub-Saharan Africans. Many Ericales species are cultivated for their showy flowers: well-known examples are azalea, rhododendron, camellia, polyanthus, cyclamen, phlox, and busy Lizzie.

Classification

The following families are typical of newer classifications. Those marked with an asterisk are recognized in the APG III system.

- Family Actinidiaceae* (kiwifruit family)
- Family Balsaminaceae* (balsam family)
- Family Clethraceae* (clethra family)
- Family Cyrillaceae* (cyrilla family)
- Family Diapensiaceae*
- Family Ebenaceae* (ebony and persimmon family)
- Family Ericaceae* (heath, rhododendron, and blueberry family)
- Family Fouquieriaceae* (Ocotillo family)
- Family Lecythidaceae* (Brazil nut family)
- Family Maesaceae → Primulaceae
- Family Marcgraviaceae*
- Family Mitrastemonaceae*
- Family Myrsinaceae (cyclamen and scarlet pimpernel family) → Primulaceae
- Family Pellicieraceae → Tetrameristaceae
- Family Pentaphylacaceae*
- Family Polemoniaceae* (phlox family)
- Family Primulaceae* (primrose and snowbell family)
- Family Roridulaceae*
- Family Sapotaceae* (sapodilla family)
- Family Sarraceniaceae* (American pitcher plant family)
- Family Sladeniaceae*
- Family Styracaceae* (silverbell family)
- Family Symplocaceae* (sapphireberry family)
- Family Ternstroemiaceae → Pentaphylacaceae
- Family Tetrameristaceae*
- Family Theaceae* (tea and camellia family)
- Family Theophrastaceae → Primulaceae

Impatiens balsamina from Family Balsaminaceae

Primula rosea from Family Primulaceae

These make up a basal group of asterids.[2] Under the Cronquist system, the Ericales included a smaller group of plants, which were placed among the Dilleniidae:

- Family Ericaceae
- Family Cyrillaceae
- Family Clethraceae
- Family Grubbiaceae
- Family Empetraceae
- Family Epacridaceae

- Family Pyrolaceae
- Family Monotropaceae

See also

- *Paradinandra*

References

[1] Robyns, W. (31). "Outline of a New System of Orders and Families of Sympetalae". *Bulletin du Jardin Botanique National Belgique* **42**: 363–372. JSTOR 3667661.

[2] Bremer, Birgitta; Kåre Bremera, Nahid Heidaria, Per Erixona, Richard G. Olmsteadb, Arne A. Anderbergc, Mari Källersjöd, Edit Barkhordarian (August 2002). "Phylogenetics of asterids based on 3 coding and 3 non-coding chloroplast DNA markers and the utility of non-coding DNA at higher taxonomic levels". *Molecular Phylogenetics and Evolution* **24** (2): 274–301. doi:10.1016/S1055-7903(02)00240-3. PMID 12144762.

- B. C. J. du Mortier (1829). *Analyse des Familles de Plantes : avec l'indication des principaux genres qui s'y rattachent*, 28. Imprimerie de J. Casterman, Tournay.
- S. Jansen, T. Watanabe, P. Caris, K. Geuten, F. Lens, N. Pyck, E. Smets (2004). The Distribution and Phylogeny of Aluminium Accumulating Plants in the Ericales. *Plant Biology (Stuttgart)* **6**, 498-505. Thieme, Stuttgart. (Available online: DOI (http://dx.doi.org/10.1055/s-2004-820980) | Abstract (http://www.thieme-connect.com/ejournals/abstract/plantbiology/doi/10.1055/s-2004-820980))
- W. S. Judd, C. S. Campbell, E. A. Kellogg, P. F. Stevens, M. J. Donoghue (2002). *Plant Systematics: A Phylogenetic Approach, 2nd edition.* pp. 425–436 (Ericales). Sinauer Associates, Sunderland, Massachusetts. ISBN 0-87893-403-0.
- E. Smets, N. Pyck (Feb 2003). Ericales (Rhododendron). In: *Nature Encyclopedia of Life Sciences*. Nature Publishing Group, London. (Available online: ELS Site (http://www.els.net))
- Arne A. Anderberg, Bertil Stahl, Mari Kallersjo (May 2000). "Maesaceae, a New Primuloid Family in the Order Ericales s.l.". *Taxon* (International Association for Plant Taxonomy (IAPT)) **49** (2): 183–187. doi:10.2307/1223834. JSTOR 1223834.

Asterids

In the APG II system (2003) for the classification of flowering plants, the name **asterids** refers to a clade (a monophyletic group).[1]

Most of the taxa belonging to this clade had been referred to the Asteridae in the Cronquist system (1981) and to the Sympetalae in earlier systems. The name **asterids** (plural, not necessarily capitalized) is presumably inspired by the earlier botanical name but in itself is intended to be the name of a clade rather than a formal ranked name, in the sense of the *ICBN*. This clade is one of the two most speciose groups of **eudicots**, the other being the **rosids**. It consists of:[1]

- clade **asterids** :

    order Cornales

    order Ericales

    clade **euasterids I**

    family Boraginaceae

    family Icacinaceae

    family Oncothecaceae

    family Vahliaceae

    order Garryales

    order Solanales

    order Gentianales

    order Lamiales

    clade **euasterids II**

    family Bruniaceae

    family Columelliaceae (+ family Desfontainiaceae)

    family Eremosynaceae

    family Escalloniaceae (+ family Tribelaceae)

    family Paracryphiaceae (+ families Sphenostemonaceae and Quintiniaceae)

    family Polyosmaceae

    order Aquifoliales

    order Apiales

    order Dipsacales

    order Asterales

Note : " + ..." = optional as a segregate of the preceding family.

External links

- Asterids [2] in Stevens, P. F. (2001 onwards). Angiosperm Phylogeny Website [3]. Version 7, May 2006.

References

[1] Angiosperm Phylogeny Group II (2003), "An update of the Angiosperm Phylogeny Group classification for the orders and families of flowering plants: APG II", Botanical Journal of the Linnean Society 141: 399–436, http://w3.ufsm.br/herb/ An%20update%20of%20the%20Angiosperm%20Phylogeny%20Group.pdf
[2] http://www.mobot.org/MOBOT/Research/APweb/orders/cornalesweb.htm#Asterids
[3] http://www.mobot.org/MOBOT/research/APweb

Eudicots

The **Eudicots, Eudicotidae** or **Eudicotyledons** are a monophyletic ground (clade or evolutionarily related group) of flowering plants that had been called *tricolpates* or *non-Magnoliid dicots* by previous authors. The botanical terms were introduced in 1991 by evolutionary botanist James A. Doyle and paleobotanist Carol L. Hotton to emphasize the later evolutionary divergence of tricolpate dicotyledons from earlier, less specialized, dicotyledons.[1] The close relationships among flowering plants with tricolpate pollen grains was initially seen in morphological studies of shared derived characters. These plants have a distinct trait in their pollen grains of exhibiting three colpi or grooves paralleling the polar axis. Later molecular evidence confirmed the genetic basis for the evolutionary relationships among flowering plants with tricolpate pollen grains and dicotyledonous traits. The term means "true dicotyledons" as it contains the majority of plants that have been considered dicotyledons and have characters of the dicotyledons. The term "eudicots" has subsequently been widely adopted in botany to refer to one of the two largest clades of angiosperms (constituting over 70% of angiosperm species), monocots being the other. The remaining angiosperms are sometimes referred to as basal angiosperms or paleodicots but these terms have not been widely or consistently adopted as the do not refer to a monophyletic group.

The other name for the eudicots is **tricolpates**, a name which refers to the grooved structure of the pollen. Members of the group have tricolpate pollen, or forms derived from it. These pollen have three or more pores set in furrows called colpi. In contrast, most of the other seed plants (that is the gymnosperms, the monocots and the paleodicots) produce monosulcate pollen, with a single pore set in a differently oriented groove called the sulcus. The name "tricolpates" is preferred by some botanists in order to avoid confusion with the dicots, a non-monophyletic group (Judd & Olmstead 2004).

A large number of familiar plants are eudicots, including many common food plants, lumber trees, and ornamentals. Some common and familiar eudicots include members of the sunflower family such as the common dandelion, the forget-me-not, cabbage and other members of the its family, apple, buttercup, maple and macadamia.

The name **eudicots** (plural) is used in the APG system, of 1998, and APG II system, of 2003, for classification of angiosperms. It is applied to a clade, a monophyletic group, which includes most of the (former) dicotyledons.

Subdivisions

The eudicots can be divided into two groups: the basal eudicots and the core eudicots.[2] Basal eudicots is an informal name for a paraphyletic group. The core eudicots are a monophyletic group.[3]

A second study has suggested that the core eudicots can be divided into two clades - Pentapetalae - comprising all core eudicots except Gunnerales - and Gunnerales.[4]

Pentapetalae can be then divided into three clades:

- (i) a "superrosid" clade consisting of Rosidae, Vitaceae and Saxifragales
- (ii) a "superasterid" clade consisting of Berberidopsidales, Santalales, Caryophyllales and Asteridae
- (iii) Dilleniaceae

Within the core eudicots, the largest groups are the "**rosids**" (core group with the prefix "eu–") and the "**asterids**" (core group with the prefix "eu–").

- **eudicots** :

    core eudicots :

    rosids :

   eurosids I

   eurosids II

    asterids :

   euasterids I

   euasterids II

In more detail, within each clade some unplaced families and orders (unplaced genera are not mentioned):

- clade **eudicots**

   family Buxaceae [+ family Didymelaceae]

   family Sabiaceae

   family Trochodendraceae [+ family Tetracentraceae]

  order Ranunculales

  order Proteales

   clade **core eudicots**

   family Aextoxicaceae

   family Berberidopsidaceae

   family Dilleniaceae

  order Gunnerales

  order Caryophyllales

  order Saxifragales

  order Santalales

   clade **rosids**

   family Aphloiaceae

   family Geissolomataceae

   family Ixerbaceae

   family Picramniaceae

   family Strassburgeriaceae

family Vitaceae

order Crossosomatales

order Geraniales

order Myrtales

clade **eurosids I**

family Zygophyllaceae [+ family Krameriaceae]

family Huaceae

order Celastrales

order Malpighiales

order Oxalidales

order Fabales

order Rosales

order Cucurbitales

order Fagales

clade **eurosids II**

family Tapisciaceae

order Brassicales

order Malvales

order Sapindales

clade **asterids**

order Cornales

order Ericales

clade **euasterids I**

family Boraginaceae

family Icacinaceae

family Oncothecaceae

family Vahliaceae

order Garryales

order Solanales

order Gentianales

order Lamiales

clade **euasterids II**

family Bruniaceae

family Columelliaceae [+ family Desfontainiaceae]

family Eremosynaceae

family Escalloniaceae

family Paracryphiaceae

family Polyosmaceae

family Sphenostemonacae

family Tribelaceae

order Aquifoliales

order Apiales

order Dipsacales

order Asterales

Note : " + ..." = optional, as a segregate of the previous family.

References

[1] Endress, Peter K. (Oct. - Dec., 2002). "Morphology and Angiosperm Systematics in the Molecular Era". *Botanical Review*. Structural Botany in Systematics: A Symposium inMemory of William C. Dickison **68** (4).

[2] Worberg A, Quandt D, Barniske A-M, Löhne C, Hilu KW, Borsch T (2007) Phylogeny of basal eudicots: insights from non-coding and rapidly evolving DNA. Organisms, Diversity and Evolution 7 (1), 55-77.

[3] Douglas E. Soltis, Pamela S. Soltis, Peter K. Endress, and Mark W. Chase. *Phylogeny and Evolution of Angiosperms*. Sinauer Associates: Sunderland, MA, USA. (2005).

[4] Moore MJ, Soltis PS, Bell CD, Burleigh JG, Soltis DE (2010) Phylogenetic analysis of 83 plastid genes further resolves the early diversification of eudicots. Proc Natl Acad Sci USA

- Doyle, J. A. & Hotton, C. L. Diversification of early angiosperm pollen in a cladistic context. pp. 169–195 in Pollen and Spores. Patterns of Diversification (eds Blackmore, S. & Barnes, S. H.) (Clarendon, Oxford, 1991).

- Walter S. Judd and Richard G. Olmstead (2004). "A survey of tricolpate (eudicot) phylogenetic relationships". *American Journal of Botany* **91** (10): 1627–1644. doi:10.3732/ajb.91.10.1627. PMID 21652313. (full text (http://www.amjbot.org/cgi/content/full/91/10/1627))

- Eudicots (http://www.mobot.org/MOBOT/Research/APweb/orders/ranunculalesweb.htm#Eudicots) in Stevens, P. F. (2001 onwards). Angiosperm Phylogeny Website (http://www.mobot.org/MOBOT/research/APweb). Version 7, May 2006.

External links

- *Eudicots* (http://www.eol.org/pages/283) at the Encyclopedia of Life
- *Eudicots* (http://www.tolweb.org/eudicots), Tree of Life Web Project
- Dicots (http://www.plantlifeforms.com/Classes186/DICOTS_DICOTYLEDONEAE_503903_186.aspx) Plant Life Forms

Plant

<table>
<tr><td colspan="2" align="center">Plants
Temporal range:
Early Cambrian to recent, but see text,</td></tr>
<tr><td colspan="2" align="center"></td></tr>
<tr><td colspan="2" align="center">Scientific classification</td></tr>
<tr><td>Domain:</td><td>Eukaryota</td></tr>
<tr><td>Unranked:</td><td>Archaeplastida</td></tr>
<tr><td>Kingdom:</td><td>Plantae
Haeckel, 1866[1]</td></tr>
<tr><td colspan="2" align="center">Divisions</td></tr>
</table>

Green algae

- Chlorophyta
- Charophyta

Land plants (embryophytes)

- **Non-vascular land plants (bryophytes)**
    - Marchantiophyta—liverworts
    - Anthocerotophyta—hornworts
    - Bryophyta—mosses
    - †Horneophytopsida
- **Vascular plants (tracheophytes)**
    - †Rhyniophyta—rhyniophytes
    - †Zosterophyllophyta—zosterophylls
    - Lycopodiophyta—clubmosses
    - †Trimerophytophyta—trimerophytes
    - Pteridophyta—ferns and horsetails
    - †Progymnospermophyta
    - **Seed plants (spermatophytes)**
        - †Pteridospermatophyta—seed ferns
        - Pinophyta—conifers
        - Cycadophyta—cycads
        - Ginkgophyta—ginkgo
        - Gnetophyta—gnetae
        - Magnoliophyta—flowering plants

†Nematophytes

Plants are living organisms belonging to the kingdom **Plantae**. Precise definitions of the kingdom vary, but as the term is used here, plants include familiar organisms such as flowering plants, conifers, ferns, mosses, and green algae, but do not include seaweeds like kelp, nor fungi and bacteria. The group is also called **green plants** or **Viridiplantae** in Latin. They obtain most of their energy from sunlight via photosynthesis using chlorophyll contained in chloroplasts, which gives them their green color. Some plants are parasitic and may not produce normal amounts of chlorophyll or photosynthesize.

Precise numbers are difficult to determine, but as of 2010, there are thought to be 300–315 thousand species of plants, of which the great majority, some 260–290 thousand, are seed plants (see the table below).[2]

The scientific study of plants is known as botany.

Definition

Plants are one of the two groups into which all living things have been traditionally divided; the other is animals. The division goes back at least as far as Aristotle (384 BC − 322 BC) who distinguished between plants which generally do not move, and animals which often are mobile to catch their food. Much later, when Linnaeus (1707–1778) created the basis of the modern system of scientific classification, these two groups became the kingdoms Vegetabilia (later Metaphyta or Plantae) and Animalia (also called Metazoa). Since then, it has become clear that the plant kingdom as originally defined included several unrelated groups, and the fungi and several groups of algae were removed to new kingdoms. However, these organisms are still often considered plants, particularly in popular contexts.

Outside of formal scientific contexts, the term "plant" implies an association with certain traits, such as being multicellular, possessing cellulose, and having the ability to carry out photosynthesis.[3] [4]

Current definitions of Plantae

When the name Plantae or plant is applied to a specific group of organisms or taxon, it usually refers to one of three concepts. From least to most inclusive, these three groupings are:

Name(s)	Scope	Description
Land plants, also known as Embryophyta or Metaphyta.	Plantae *sensu strictissimo*	This group includes the liverworts, hornworts, mosses, and vascular plants, as well as fossil plants similar to these surviving groups.
Green plants - also known as **Viridiplantae**, **Viridiphyta** or **Chlorobionta**	Plantae *sensu stricto*	This group includes the land plants plus various groups of green algae, including stoneworts. The names given to these groups vary considerably as of July 2011. Viridiplantae encompass a group of organisms that possess chlorophyll *a* and *b*, have plastids that are bound by only two membranes, are capable of storing starch, and have cellulose in their cell walls. It is this clade which is mainly the subject of this article.
Archaeplastida, Plastida or Primoplantae	Plantae *sensu lato*	This group comprises the green plants above plus Rhodophyta (red algae) and Glaucophyta (glaucophyte algae). This clade includes the organisms that eons ago acquired their chloroplasts directly by engulfing cyanobacteria.

Another way of looking at the relationships between the different groups which have been called "plants" is through a cladogram, which shows their evolutionary relationships. The evolutionary history of plants is not yet completely settled, but one accepted relationship between the three groups described above is shown below.[5] Those which have been called "plants" are in bold.

Glaucophyta (glaucophyte algae)

Rhodophyta (red algae)

Chlorophyta (part of green algae)

streptophyte algae (part of green algae)

Archaeplastida

Viridiplantae

Streptophyta

Charales (stoneworts, often included
in green algae)

land plants or embryophytes

The way in which the groups of green algae are combined and named varies considerably between authors.

Many of the classification controversies involve organisms that are rarely encountered and are of minimal apparent economic significance, but are crucial in developing an understanding of the evolution of modern flora.

Algae

Algae comprise several different groups of organisms which produce energy through photosynthesis and for that reason have been included in the plant kingdom in the past. Most conspicuous among the algae are the seaweeds, multicellular algae that may roughly resemble land plants, but are classified among the brown, red and green algae. Each of these algal groups also includes various microscopic and single-celled organisms. There is good evidence that some of these algal groups arose independently from separate non-photosynthetic ancestors, with the result that many groups of algae are no longer classified within the plant kingdom as it is defined here.[6] [7]

The Viridiplantae, the green plants – green algae and land plants – form a clade, a group consisting of all the descendants of a common ancestor. With a few exceptions among the green algae, all green plants have many features in common, including cell walls containing cellulose, chloroplasts containing chlorophylls *a* and *b*, and food stores in the form of starch. They undergo closed mitosis without centrioles, and typically have mitochondria with flat cristae. The chloroplasts of green plants are surrounded by two membranes, suggesting they originated directly from endosymbiotic cyanobacteria.

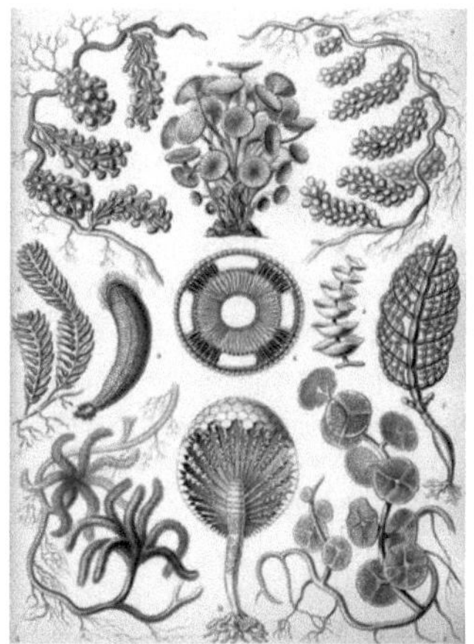

Green algae from Ernst Haeckel's *Kunstformen der Natur*, 1904.

Two additional groups, the Rhodophyta (red algae) and Glaucophyta (glaucophyte algae), also have chloroplasts which appear to be derived directly from endosymbiotic cyanobacteria, although they differ in the pigments which are used in photosynthesis and so are different in colour. All three groups together are generally believed to have a single common origin, and so are classified together in the taxon Archaeplastida, whose name implies that the chloroplasts or plastids of all the members of the taxon were derived from a single ancient endosymbiotic event. This is the broadest modern definition of the plants.

In contrast, most other algae (e.g. heterokonts, haptophytes, dinoflagellates, and euglenids) not only have different pigments but also have chloroplasts with three or four surrounding membranes. They are not close relatives of the Archaeplastida, presumably having acquired chloroplasts separately from ingested or symbiotic green and red algae. They are thus not included in even the broadest modern definition of the plant kingdom, although they were in the past.

The green plants or Viridiplantae were traditionally divided into the green algae (including the stoneworts) and the land plants. However, it is now known that the land plants evolved from within a group of green algae, so that the green algae by themselves are a paraphyletic group, i.e. a group which excludes some of the descendants of a common ancestor. Paraphyletic groups are generally avoided in modern classifications, so that in recent treatments the Viridiplantae have been divided into two clades, the Chlorophyta and the Streptophyta (or Charophyta).[8] [9]

The Chlorophyta (a name that has also been used for *all* green algae) are the sister group to the group from which the land plants evolved. There are about 4,300 species[10] of mainly marine organisms, both unicellular and multicellular. The latter include the sea lettuce, *Ulva*.

The other group within the Viridiplantae are the mainly freshwater or terrestrial Streptophyta (or Charophyta), which consist of several groups of green algae plus the stoneworts and land plants. (The names have been used differently, e.g. Streptophyta to mean the group which excludes the land plants and Charophyta for the stoneworts alone or the stoneworts plus the land plants.) Streptophyte algae are either unicellular or form multicellular filaments, branched or unbranched.[9] The genus *Spirogyra* is a filamentous streptophyte alga familiar to many, as it is often used in teaching and is one of the organisms responsible for the algal "scum" which pond-owners so dislike. The freshwater stoneworts strongly resemble land plants and are believed to be their closest relatives. Growing underwater, they consist of a central stalk with whorls of branchlets, giving them a superficial resemblance to horsetails, species of the genus *Equisetum*, which are true land plants.

Fungi

The classification of fungi has been controversial until quite recently in the history of biology. Linnaeus' original classification placed the fungi within the Plantae, since they were unquestionably not animals or minerals and these were the only other alternatives. With later developments in microbiology, in the 19th century Ernst Haeckel felt that another kingdom was required to classify newly discovered micro-organisms. The introduction of the new kingdom Protista in addition to Plantae and Animalia, led to uncertainty as to whether fungi truly were best placed in the Plantae or whether they ought to be reclassified as protists. Haeckel himself found it difficult to decide and it was not until 1969 that a solution was found whereby Robert Whittaker proposed the creation of the kingdom Fungi. Molecular evidence has since shown that the last common ancestor (concestor) of the Fungi was probably more similar to that of the Animalia than of any other kingdom, including the Plantae.

Whittaker's original reclassification was based on the fundamental difference in nutrition between the Fungi and the Plantae. Unlike plants, which generally gain carbon through photosynthesis, and so are called autotrophic phototrophs, fungi generally obtain carbon by breaking down and absorbing surrounding materials, and so are called heterotrophic saprotrophs. In addition, the substructure of multicellular fungi is different from that of plants, taking the form of many chitinous microscopic strands called hyphae, which may be further subdivided into cells or may form a syncytium containing many eukaryotic nuclei. Fruiting bodies, of which mushrooms are most familiar example, are the reproductive structures of fungi, and are unlike any structures produced by plants.

Diversity

The table below shows some species count estimates of different green plant (Viridiplantae) divisions. It suggests there are about 300,000 species of living Viridiplantae, of which 85-90% are flowering plants. (Note: as these are from different sources and different dates, they are not necessarily comparable, and like all species counts, are subject to a degree of uncertainty in some cases.)

Diversity of living green plant (Viridiplantae) divisions

Informal group	Division name	Common name	No. of living species	Approximate No. in informal group
Green algae	Chlorophyta	green algae (chlorophytes)	3,800 [11] – 4,300 [12]	8,500 (6,600 - 10,300)
	Charophyta	green algae (e.g. desmids & stoneworts)	2,800; [13] 4,000-6,000 [14]	
Bryophytes	Marchantiophyta	liverworts	6,000-8,000 [15]	19,000 (18,100 - 20,200)
	Anthocerotophyta	hornworts	100-200 [16]	
	Bryophyta	mosses	12,000 [17]	
Pteridophytes	Lycopodiophyta	club mosses	1,200 [7]	12,000 (12,200)
	Pteridophyta	ferns, whisk ferns & horsetails	11,000 [7]	
Seed plants	Cycadophyta	cycads	160 [18]	260,000 (259,511)
	Ginkgophyta	ginkgo	1 [19]	
	Pinophyta	conifers	630 [7]	
	Gnetophyta	gnetophytes	70 [7]	
	Magnoliophyta	flowering plants	258,650 [20]	

The naming of plants is governed by the International Code of Botanical Nomenclature and International Code of Nomenclature for Cultivated Plants (see cultivated plant taxonomy).

Evolution

Further information: Evolutionary history of plants

The evolution of plants has resulted in increasing levels of complexity, from the earliest algal mats, through bryophytes, lycopods, ferns to the complex gymnosperms and angiosperms of today. The groups which appeared earlier continue to thrive, especially in the environments in which they evolved.

Evidence suggests that an algal scum formed on the land 1200 [21] million years ago, but it was not until the Ordovician Period, around 450.0 [22] million years ago, that land plants appeared.[23] However, new evidence from the study of carbon isotope ratios in Precambrian rocks has suggested that complex photosynthetic plants developed on the earth over 1000 m.y.a.[24] These began to diversify in the late Silurian Period, around 420 [25] million years ago, and the fruits of their diversification are displayed in remarkable detail in an early Devonian fossil assemblage from the Rhynie chert. This chert preserved early plants in cellular detail, petrified in volcanic springs. By the middle of the Devonian Period most of the features recognised in plants today are present, including roots, leaves and secondary wood, and by late Devonian times seeds had evolved.[26] Late Devonian plants had thereby reached a degree of sophistication that allowed them to form forests of tall trees. Evolutionary innovation continued after the Devonian period. Most plant groups were relatively unscathed by the Permo-Triassic extinction event, although the structures of communities changed. This may have set the scene for the evolution of flowering plants in the Triassic (~200 [27] million years ago), which exploded in the Cretaceous and Tertiary. The latest major group of plants to evolve were the grasses, which became important in the mid Tertiary, from around 40 [28] million years ago. The grasses, as well as many other groups, evolved new mechanisms of metabolism to survive the low CO_2 and warm, dry conditions of the tropics over the last 10 [29] million years.

A proposed phylogenetic tree of Plantae, after Kenrick and Crane,[30] is as follows, with modification to the Pteridophyta from Smith et al.[31] The Prasinophyceae may be a paraphyletic basal group to all green plants.

Prasinophyceae (micromonads)

Lignophytia

Spermatophytes (seed plants)

Progymnospermophyta †

Euphyllophytina

Pteridopsida (true ferns)

Marattiopsida

Eutracheophytes

Tracheophytes

Pteridophyta

Equisetopsida (horsetails)

Psilotopsida (whisk ferns & adders'-tongues)

Polysporangiates

Cladoxylopsida †

Stomatophytes

Embryophytes

eptobionta

Lycophytina

Lycopodiophyta

Zosterophyllophyta †

Rhyniophyta †

Aglaophyton †

Horneophytopsida †

Bryophyta (mosses)

Anthocerotophyta (hornworts)

Marchantiophyta (liverworts)

Charophyta

	Trebouxiophyceae
	(Pleurastrophyceae)
Chlorophyta	Chlorophyceae
	Ulvophyceae

Embryophytes

The plants that are likely most familiar to us are the multicellular land plants, called embryophytes. They include the vascular plants, plants with full systems of leaves, stems, and roots. They also include a few of their close relatives, often called *bryophytes*, of which mosses and liverworts are the most common.

All of these plants have eukaryotic cells with cell walls composed of cellulose, and most obtain their energy through photosynthesis, using light and carbon dioxide to synthesize food. About three hundred plant species do not photosynthesize but are parasites on other species of photosynthetic plants. Plants are distinguished from green algae, which represent a mode of photosynthetic life similar to the kind modern plants are believed to have evolved from, by having specialized reproductive organs protected by non-reproductive tissues.

Dicksonia antarctica, a species of tree fern

Bryophytes first appeared during the early Paleozoic. They can only survive where moisture is available for significant periods, although some species are desiccation tolerant. Most species of bryophyte remain small throughout their life-cycle. This involves an alternation between two generations: a haploid stage, called the gametophyte, and a diploid stage, called the sporophyte. The sporophyte is short-lived and remains dependent on its parent gametophyte.

Vascular plants first appeared during the Silurian period, and by the Devonian had diversified and spread into many different land environments. They have a number of adaptations that allowed them to overcome the limitations of the bryophytes. These include a cuticle resistant to desiccation, and vascular tissues which transport water throughout the organism. In most the sporophyte acts as a separate individual, while the gametophyte remains small.

The first primitive seed plants, Pteridosperms (seed ferns) and Cordaites, both groups now extinct, appeared in the late Devonian and diversified through the Carboniferous, with further evolution through the Permian and Triassic periods. In these the gametophyte stage is completely reduced, and the sporophyte begins life inside an enclosure called a seed, which develops while on the parent plant, and with fertilisation by means of pollen grains. Whereas other vascular plants, such as ferns, reproduce by means of spores and so need moisture to develop, some seed plants can survive and reproduce in extremely arid conditions.

Early seed plants are referred to as gymnosperms (naked seeds), as the seed embryo is not enclosed in a protective structure at pollination, with the pollen landing directly on the embryo. Four surviving groups remain widespread now, particularly the conifers, which are dominant trees in several biomes. The angiosperms, comprising the flowering plants, were the last major group of plants to appear, emerging from within the gymnosperms during the Jurassic and diversifying rapidly during the Cretaceous. These differ in that the seed embryo (angiosperm) is enclosed, so the pollen has to grow a tube to penetrate the protective seed coat; they are the predominant group of flora in most biomes today.

Fossils

Plant fossils include roots, wood, leaves, seeds, fruit, pollen, spores, phytoliths, and amber (the fossilized resin produced by some plants). Fossil land plants are recorded in terrestrial, lacustrine, fluvial and nearshore marine sediments. Pollen, spores and algae (dinoflagellates and acritarchs) are used for dating sedimentary rock sequences. The remains of fossil plants are not as common as fossil animals, although plant fossils are locally abundant in many regions worldwide.

The earliest fossils clearly assignable to Kingdom Plantae are fossil green algae from the Cambrian. These fossils resemble calcified multicellular members of the Dasycladales. Earlier Precambrian fossils are known which resemble single-cell green algae, but definitive identity with that group of algae is uncertain.

The oldest known fossils of embryophytes date from the Ordovician, though such fossils are fragmentary. By the Silurian, fossils of whole plants are preserved, including the lycophyte *Baragwanathia longifolia*. From the Devonian, detailed fossils of rhyniophytes have been found. Early fossils of these ancient plants show the individual cells within the plant tissue. The Devonian period also saw the

A petrified log in Petrified Forest National Park.

evolution of what many believe to be the first modern tree, *Archaeopteris*. This fern-like tree combined a woody trunk with the fronds of a fern, but produced no seeds.

The Coal measures are a major source of Paleozoic plant fossils, with many groups of plants in existence at this time. The spoil heaps of coal mines are the best places to collect; coal itself is the remains of fossilised plants, though structural detail of the plant fossils is rarely visible in coal. In the Fossil Forest at Victoria Park in Glasgow, Scotland, the stumps of *Lepidodendron* trees are found in their original growth positions.

The fossilized remains of conifer and angiosperm roots, stems and branches may be locally abundant in lake and inshore sedimentary rocks from the Mesozoic and Cenozoic eras. Sequoia and its allies, magnolia, oak, and palms are often found.

Petrified wood is common in some parts of the world, and is most frequently found in arid or desert areas where it is more readily exposed by erosion. Petrified wood is often heavily silicified (the organic material replaced by silicon dioxide), and the impregnated tissue is often preserved in fine detail. Such specimens may be cut and polished using lapidary equipment. Fossil forests of petrified wood have been found in all continents.

Fossils of seed ferns such as *Glossopteris* are widely distributed throughout several continents of the Southern Hemisphere, a fact that gave support to Alfred Wegener's early ideas regarding Continental drift theory.

Structure, growth, and development

Further information: Plant morphology

Most of the solid material in a plant is taken from the atmosphere. Through a process known as photosynthesis, most plants use the energy in sunlight to convert carbon dioxide from the atmosphere, plus water, into simple sugars. Parasitic plants, on the other hand, use the resources of its host to grow. These sugars are then used as building blocks and form the main structural component of the plant. Chlorophyll, a green-colored, magnesium-containing pigment is essential to this process; it is generally present in plant leaves, and often in other plant parts as well.

Plants usually rely on soil primarily for support and water (in quantitative terms), but also obtain compounds of nitrogen, phosphorus, and other crucial elemental nutrients. Epiphytic and lithophytic plants often depend on rainwater or other sources for nutrients and carnivorous plants supplement their nutrient requirements with insect prey that they capture. For the majority of plants to grow successfully they also require oxygen in the atmosphere and around their roots for respiration. However, some plants grow as submerged aquatics, using oxygen dissolved in the surrounding water, and a few specialized vascular plants, such as mangroves, can grow with their roots in anoxic conditions.

Factors affecting growth

The genotype of a plant affects its growth. For example, selected varieties of wheat grow rapidly, maturing within 110 days, whereas others, in the same environmental conditions, grow more slowly and mature within 155 days.[32]

Growth is also determined by environmental factors, such as temperature, available water, available light, and available nutrients in the soil. Any change in the availability of these external conditions will be reflected in the plants growth.

The leaf is usually the primary site of photosynthesis in plants.

Biotic factors are also capable of affecting plant growth. Plants compete with other plants for space, water, light and nutrients. Plants can be so crowded that no single individual produces normal growth, causing etiolation and chlorosis. Optimal plant growth can be hampered by grazing animals, suboptimal soil composition, lack of mycorrhizal fungi, and attacks by insects or plant diseases, including those caused by bacteria, fungi, viruses, and nematodes.[32]

Simple plants like algae may have short life spans as individuals, but their populations are commonly seasonal. Other plants may be organized according to their seasonal growth pattern: annual plants live and reproduce within one growing season, biennial plants live for two growing seasons and usually reproduce in second year, and perennial plants live for many growing seasons and continue to reproduce once they are mature. These designations often depend on climate and other environmental factors; plants that are annual in alpine or temperate regions can be biennial or perennial in warmer climates. Among the

There is no photosynthesis in deciduous leaves in autumn.

vascular plants, perennials include both evergreens that keep their leaves the entire year, and deciduous plants which lose their leaves for some part of it. In temperate and boreal climates, they generally lose their leaves during the winter; many tropical plants lose their leaves during the dry season.

The growth rate of plants is extremely variable. Some mosses grow less than 0.001 millimeters per hour (mm/h), while most trees grow 0.025-0.250 mm/h. Some climbing species, such as kudzu, which do not need to produce thick supportive tissue, may grow up to 12.5 mm/h.

Plants protect themselves from frost and dehydration stress with antifreeze proteins, heat-shock proteins and sugars (sucrose is common). LEA (Late Embryogenesis Abundant) protein expression is induced by stresses and protects other proteins from aggregation as a result of desiccation and freezing.[33]

Dried dead plants

Plant cell

Plant cells are typically distinguished by their large water-filled central vacuole, chloroplasts, and rigid cell walls that are made up of cellulose, hemicellulose, and pectin. Cell division is also characterized by the development of a phragmoplast for the construction of a cell plate in the late stages of cytokinesis. Just as in animals, plant cells differentiate and develop into multiple cell types. Totipotent meristematic cells can differentiate into vascular, storage, protective (e.g. epidermal layer), or reproductive tissues, with more primitive plants lacking some tissue types.[34]

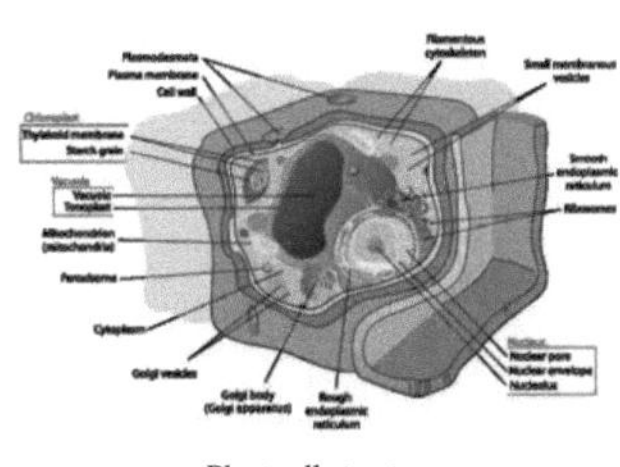

Plant cell structure

Physiology

Photosynthesis

Plants are photosynthetic, which means that they manufacture their own food molecules using energy obtained from light. The primary mechanism plants have for capturing light energy is the pigment chlorophyll. All green plants contain two forms of chlorophyll, chlorophyll *a* and chlorophyll *b*. The latter of these pigments is not found in red or brown algae.

Immune system

By means of cells that behave like nerves, plants receive and distribute within their systems information about incident light intensity and quality. Incident light which stimulates a chemical reaction in one leaf, will cause a chain reaction of signals to the entire plant via a type of cell termed a *bundle sheath cell*. Researchers from the Warsaw University of Life Sciences in Poland, found that plants have a specific memory for varying light conditions which prepares their immune systems against seasonal pathogens.[35] Plants use pattern-recognition receptors to recognize conserved microbial signatures. This recognition triggers an immune response. The first plant receptors of conserved microbial signatures were identified in rice (XA21, 1995)[36] and in *Arabidopsis* (FLS2, 2000).[37] Plants also carry immune receptors that recognize highly variable pathogen effectors. These include the NBS-LRR class of proteins.

Internal distribution

Vascular plants differ from other plants in that they transport nutrients between different parts through specialized structures, called xylem and phloem. They also have roots for taking up water and minerals. The xylem moves water and minerals from the root to the rest of the plant, and the phloem provides the roots with sugars and other nutrient produced by the leaves.[34]

Ecology

The photosynthesis conducted by land plants and algae is the ultimate source of energy and organic material in nearly all ecosystems. Photosynthesis radically changed the composition of the early Earth's atmosphere, which as a result is now 21% oxygen. Animals and most other organisms are aerobic, relying on oxygen; those that do not are confined to relatively rare anaerobic environments. Plants are the primary producers in most terrestrial ecosystems and form the basis of the food web in those ecosystems. Many animals rely on plants for shelter as well as oxygen and food.

Land plants are key components of the water cycle and several other biogeochemical cycles. Some plants have coevolved with nitrogen fixing bacteria, making plants an important part of the nitrogen cycle. Plant roots play an essential role in soil development and prevention of soil erosion.

Distribution

Plants are distributed worldwide in varying numbers. While they inhabit a multitude of biomes and ecoregions, few can be found beyond the tundras at the northernmost regions of continental shelves. At the southern extremes, plants have adapted tenaciously to the prevailing conditions. (See Antarctic flora.)

Plants are often the dominant physical and structural component of habitats where they occur. Many of the Earth's biomes are named for the type of vegetation because plants are the dominant organisms in those biomes, such as grasslands and forests.

Ecological relationships

Numerous animals have coevolved with plants. Many animals pollinate flowers in exchange for food in the form of pollen or nectar. Many animals disperse seeds, often by eating fruit and passing the seeds in their feces. Myrmecophytes are plants that have coevolved with ants. The plant provides a home, and sometimes food, for the ants. In exchange, the ants defend the plant from herbivores and sometimes competing plants. Ant wastes provide organic fertilizer.

The majority of plant species have various kinds of fungi associated with their root systems in a kind of mutualistic symbiosis known as mycorrhiza. The fungi help the plants gain water and mineral nutrients from the soil, while the plant gives the fungi carbohydrates manufactured in photosynthesis. Some plants serve as homes for endophytic fungi that protect the plant from herbivores by producing toxins. The fungal endophyte, *Neotyphodium coenophialum*, in tall fescue (*Festuca arundinacea*) does tremendous economic damage to the cattle industry in the U.S.

The Venus flytrap, a species of carnivorous plant.

Various forms of parasitism are also fairly common among plants, from the semi-parasitic mistletoe that merely takes some nutrients from its host, but still has photosynthetic leaves, to the fully parasitic broomrape and toothwort that acquire all their nutrients through connections to the roots of other plants, and so have no chlorophyll. Some plants, known as myco-heterotrophs, parasitize mycorrhizal fungi, and hence act as epiparasites on other plants.

Many plants are epiphytes, meaning they grow on other plants, usually trees, without parasitizing them. Epiphytes may indirectly harm their host plant by intercepting mineral nutrients and light that the host would otherwise receive. The weight of large numbers of epiphytes may break tree limbs. Hemiepiphytes like the strangler fig begin as epiphytes but eventually set their own roots and overpower and kill their host. Many orchids, bromeliads, ferns and mosses often grow as epiphytes. Bromeliad epiphytes accumulate water in leaf axils to form phytotelmata, complex aquatic food webs.[38]

Approximately 630 plants are carnivorous, such as the Venus Flytrap (*Dionaea muscipula*) and sundew (*Drosera species*). They trap small animals and digest them to obtain mineral nutrients, especially nitrogen and phosphorus.[39]

Importance

The study of plant uses by people is termed economic botany or ethnobotany; some consider economic botany to focus on modern cultivated plants, while ethnobotany focuses on indigenous plants cultivated and used by native peoples. Human cultivation of plants is part of agriculture, which is the basis of human civilization. Plant agriculture is subdivided into agronomy, horticulture and forestry.

Food

Much of human nutrition depends on land plants, either directly or indirectly.

Human nutrition depends to a large extent on cereals, especially maize (or corn), wheat and rice. Other staple crops include potato, cassava, and legumes. Human food also includes vegetables, spices, and certain fruits, nuts, herbs, and edible flowers.

Beverages produced from plants include coffee, tea, wine, beer and alcohol.

Sugar is obtained mainly from sugar cane and sugar beet.

Cooking oils and margarine come from maize, soybean, rapeseed, safflower, sunflower, olive and others.

Food additives include gum arabic, guar gum, locust bean gum, starch and pectin.

Livestock animals including cows, pigs, sheep, and goats are all herbivores; and feed primarily or entirely on cereal plants, particularly grasses.

Potato plant. Potatoes spread to the rest of the world after European contact with the Americas in the late 15th and early 16th centuries and have since become an important field crop.

Nonfood products

Wood is used for buildings, furniture, paper, cardboard, musical instruments and sports equipment. Cloth is often made from cotton, flax or synthetic fibers derived from cellulose, such as rayon and

Timber in storage for later processing at a sawmill.

acetate. Renewable fuels from plants include firewood, peat and many other biofuels. Coal and petroleum are fossil fuels derived from plants. Medicines derived from plants include aspirin, taxol, morphine, quinine, reserpine, colchicine, digitalis and vincristine. There are hundreds of herbal supplements such as ginkgo, Echinacea, feverfew, and Saint John's wort. Pesticides derived from plants include nicotine, rotenone, strychnine and pyrethrins. Drugs obtained from plants include opium, cocaine and marijuana. Poisons from plants include ricin, hemlock and curare. Plants are the source of many natural products such as fibers, essential oils, natural dyes, pigments, waxes, tannins, latex, gums, resins, alkaloids, amber and cork. Products derived from plants include soaps, paints, shampoos, perfumes, cosmetics, turpentine, rubber, varnish, lubricants, linoleum, plastics, inks, chewing gum and hemp rope. Plants are also a primary source of basic chemicals for the industrial synthesis of a vast array of organic chemicals. These chemicals are used in a vast variety of studies and experiments.

A section of a Yew branch showing 27 annual growth rings, pale sapwood and dark heartwood, and pith (centre dark spot). The dark radial lines are longitudinal sections of small branches which became included by growth of the tree.

Aesthetic uses

Thousands of plant species are cultivated for aesthetic purposes as well as to provide shade, modify temperatures, reduce wind, abate noise, provide privacy, and prevent soil erosion. People use cut flowers, dried flowers and houseplants indoors or in greenhouses. In outdoor gardens, lawn grasses, shade trees, ornamental trees, shrubs, vines, herbaceous perennials and bedding plants are used. Images of plants are often used in art, architecture, humor, language, and photography and on textiles, money, stamps, flags and coats of arms. Living plant art forms include topiary, bonsai, ikebana and espalier. Ornamental plants have sometimes changed the course of history, as in tulipomania. Plants are the basis of a multi-billion dollar per year tourism industry which includes travel to arboretums, botanical gardens, historic gardens, national parks, tulip festivals, rainforests, forests with colorful autumn leaves and the National Cherry Blossom Festival. Venus Flytrap, sensitive plant and resurrection plant are examples of plants sold as novelties.

Scientific and cultural uses

Tree rings are an important method of dating in archeology and serve as a record of past climates. Basic biological research has often been done with plants, such as the pea plants used to derive Gregor Mendel's laws of genetics. Space stations or space colonies may one day rely on plants for life support. Plants are used as national and state emblems, including state trees and state flowers. Ancient trees are revered and many are famous. Numerous world records are held by plants. Plants are often used as memorials, gifts and to mark special occasions such as births, deaths, weddings and holidays. Plants figure prominently in mythology, religion and literature. The field of ethnobotany studies plant use by indigenous cultures which helps to conserve endangered species as well as discover new medicinal plants. Gardening is the most popular leisure activity in the U.S. Working with plants or horticulture therapy is beneficial for rehabilitating people with disabilities. Certain plants contain psychotropic chemicals which are extracted and ingested, including tobacco, cannabis (marijuana), and opium.

Negative effects

Weeds are plants that grow where people do not want them. People have spread plants beyond their native ranges and some of these introduced plants become invasive, damaging existing ecosystems by displacing native species. Invasive plants cause billions of dollars in crop losses annually by displacing crop plants, they increase the cost of production and the use of chemical means to control them affects the environment.

Plants may cause harm to animals, including people. Plants that produce windblown pollen invoke allergic reactions in people who suffer from hay fever. A wide variety of plants are poisonous. Toxalbumins are plant poisons fatal to most mammals and act as a serious deterrent to consumption. Several plants cause skin irritations when touched, such as poison ivy. Certain plants contain psychotropic chemicals, which are extracted and ingested or smoked, including tobacco, cannabis (marijuana), cocaine and opium. Smoking causes damage to health or even death, while some drugs may also be harmful or fatal to people.[40] [41] Both illegal and legal drugs derived from plants may have negative effects on the economy, affecting worker productivity and law enforcement costs.[42] [43] Some plants cause allergic reactions when ingested, while other plants cause food intolerances that negatively affect health.

See also

- The Plant List
- Biosphere
- Leaf sensor
- Plant defense against herbivory
- Plant perception (paranormal)
- Plant perception (physiology)
- Plant identification
- Rapid plant movement
- Biological exponential growth
- Germination

References

[1] Haeckel G (1866). *Generale Morphologie der Organismen*. Berlin: Verlag von Georg Reimer. pp. vol.1: i–xxxii, 1–574, pls I–II; vol. 2: i–clx, 1–462, pls I–VIII.

[2] http://www.iucnredlist.org/documents/summarystatistics/2010_1RL_Stats_Table_1.pdf

[3] "plant[2 (http://www.merriam-webster.com/dictionary/plant[2]) - Definition from the Merriam-Webster Online Dictionary"]. . Retrieved 2009-03-25.

[4] "plant (life form) -- Britannica Online Encyclopedia" (http://www.britannica.com/EBchecked/topic/463192/plant). . Retrieved 2009-03-25.

[5] Based on Rogozin, I.B.; Basu, M.K.; Csűrös, M. & Koonin, E.V. (2009), "Analysis of Rare Genomic Changes Does Not Support the Unikont–Bikont Phylogeny and Suggests Cyanobacterial Symbiosis as the Point of Primary Radiation of Eukaryotes", *Genome Biology and Evolution* 1: 99–113, doi:10.1093/gbe/evp011, PMC 2817406, PMID 20333181 and Becker, B. & Marin, B. (2009), "Streptophyte algae and the origin of embryophytes", *Annals of Botany* 103 (7): 999–1004, doi:10.1093/aob/mcp044, PMC 2707909, PMID 19273476; see also the slightly different cladogram in Lewis, Louise A. & McCourt, R.M. (2004), "Green algae and the origin of land plants", *Am. J. Bot.* 91 (10): 1535–1556, doi:10.3732/ajb.91.10.1535, PMID 21652308.

[6] Margulis, L. (1974). "Five-kingdom classification and the origin and evolution of cells". *Evolutionary Biology* 7: 45–78.

[7] Raven, Peter H., Ray F. Evert, & Susan E. Eichhorn, 2005. *Biology of Plants*, 7th edition. (New York: W. H. Freeman and Company). ISBN 0-7167-1007-2.

[8] Lewis, Louise A. & McCourt, R.M. (2004), "Green algae and the origin of land plants", *Am. J. Bot.* 91 (10): 1535–1556, doi:10.3732/ajb.91.10.1535, PMID 21652308

[9] Becker, B. & Marin, B. (2009), "Streptophyte algae and the origin of embryophytes", *Annals of Botany* 103 (7): 999–1004, doi:10.1093/aob/mcp044, PMC 2707909, PMID 19273476

[10] Guiry, M.D. & Guiry, G.M. (2007). "Phylum: Chlorophyta taxonomy browser" (http://www.algaebase.org/browse/taxonomy/?id=4307). *AlgaeBase version 4.2* World-wide electronic publication, National University of Ireland, Galway. . Retrieved 2007-09-23.

[11] Van den Hoek, C., D. G. Mann, & H. M. Jahns, 1995. *Algae: An Introduction to Phycology.* pages 343, 350, 392, 413, 425, 439, & 448 (Cambridge: Cambridge University Press). ISBN 0-521-30419-9

[12] Guiry, M.D. & Guiry, G.M. (2011), *AlgaeBase : Chlorophyta* (http://www.algaebase.org/browse/taxonomy/?searching=true& gettaxon=Chlorophyta), World-wide electronic publication, National University of Ireland, Galway, , retrieved 2011-07-26

[13] Guiry, M.D. & Guiry, G.M. (2011), *AlgaeBase : Charophyta* (http://www.algaebase.org/browse/taxonomy/?searching=true& gettaxon=Charophyta), World-wide electronic publication, National University of Ireland, Galway, , retrieved 2011-07-26

[14] Van den Hoek, C., D. G. Mann, & H. M. Jahns, 1995. *Algae: An Introduction to Phycology.* pages 457, 463, & 476. (Cambridge: Cambridge University Press). ISBN 0-521-30419-9

[15] Crandall-Stotler, Barbara. & Stotler, Raymond E., 2000. "Morphology and classification of the Marchantiophyta". page 21 *in* A. Jonathan Shaw & Bernard Goffinet (Eds.), *Bryophyte Biology.* (Cambridge: Cambridge University Press). ISBN 0-521-66097-1

[16] Schuster, Rudolf M., *The Hepaticae and Anthocerotae of North America*, volume VI, pages 712-713. (Chicago: Field Museum of Natural History, 1992). ISBN 0-914-86821-7.

[17] Goffinet, Bernard; William R. Buck (2004). "Systematics of the Bryophyta (Mosses): From molecules to a revised classification". *Monographs in Systematic Botany* (Missouri Botanical Garden Press) **98**: 205–239.

[18] Gifford, Ernest M. & Adriance S. Foster, 1988. *Morphology and Evolution of Vascular Plants*, 3rd edition, page 358. (New York: W. H. Freeman and Company). ISBN 0-7167-1946-0.

[19] Taylor, Thomas N. & Edith L. Taylor, 1993. *The Biology and Evolution of Fossil Plants*, page 636. (New Jersey: Prentice-Hall). ISBN 0-13-651589-4.

[20] International Union for Conservation of Nature and Natural Resources, 2006. *IUCN Red List of Threatened Species:Summary Statistics* (http://www.iucnredlist.org/)

[21] http://toolserver.org/~verisimilus/Timeline/Timeline.php?Ma=1200

[22] http://toolserver.org/~verisimilus/Timeline/Timeline.php?Ma=450

[23] "The oldest fossils reveal evolution of non-vascular plants by the middle to late Ordovician Period (~450-440 m.y.a.) on the basis of fossil spores" Transition of plants to land (http://www.clas.ufl.edu/users/pciesiel/gly3150/plant.html)

[24] "The apparent dominance of eukaryotes in non-marine settings by 1 Gyr ago indicates that eukaryotic evolution on land may have commenced far earlier than previously thought." Earth's earliest non-marine eukaryotes (http://www.nature.com/nature/journal/vaop/ncurrent/full/nature09943.html)

[25] http://toolserver.org/~verisimilus/Timeline/Timeline.php?Ma=420

[26] Rothwell, G. W.; Scheckler, S. E.; Gillespie, W. H. (1989). "*Elkinsia* gen. nov., a Late Devonian gymnosperm with cupulate ovules". *Botanical Gazette* **150** (2): 170–189. doi:10.1086/337763.

[27] http://toolserver.org/~verisimilus/Timeline/Timeline.php?Ma=200

[28] http://toolserver.org/~verisimilus/Timeline/Timeline.php?Ma=40

[29] http://toolserver.org/~verisimilus/Timeline/Timeline.php?Ma=10

[30] Kenrick, Paul & Peter R. Crane. 1997. *The Origin and Early Diversification of Land Plants: A Cladistic Study.* (Washington, D.C.: Smithsonian Institution Press). ISBN 1-56098-730-8.

[31] Smith Alan R., Pryer Kathleen M., Schuettpelz E., Korall P., Schneider H., Wolf Paul G. (2006). "A classification for extant ferns" (http:// www.pryerlab.net/publication/fichier749.pdf) (PDF). *Taxon* **55** (3): 705–731. doi:10.2307/25065646. .

[32] Robbins, W.W., Weier, T.E., *et al.*, *Botany:Plant Science*, 3rd edition , Wiley International, New York, 1965.

[33] Goyal, K., Walton, L. J., & Tunnacliffe, A. (2005). "LEA proteins prevent protein aggregation due to water stress" (http://www. webcitation.org/5il9QhYT0). *Biochemical Journal* **388** (Part 1): 151–157. doi:10.1042/BJ20041931. PMC 1186703. PMID 15631617. Archived from the original (http://www.biochemj.org/bj/388/0151/bj3880151.htm) on 2009-08-03. .

[34] Campbell, Reece, *Biology*, 7th edition, Pearson/Benjamin Cummings, 2005.

[35] BBC Report (http://www.bbc.co.uk/news/10598926)

[36] Song, W.Y. et al. (1995). "A receptor kinase-like protein encoded by the rice disease resistance gene, XA21". *Science* **270** (5243): 1804–1806. doi:10.1126/science.270.5243.1804. PMID 8525370.

[37] Gomez-Gomez, L. et al. (2000). "FLS2: an LRR receptor-like kinase involved in the perception of the bacterial elicitor flagellin in *Arabidopsis*". *Molecular Cell* **5** (6): 1003–1011. doi:10.1016/S1097-2765(00)80265-8. PMID 10911994.

[38] Howard Frank, Bromeliad Phytotelmata (http://entomology.ifas.ufl.edu/frank/bromeliadbiota/bromfit.htm), October 2000

[39] Barthlott, W., S. Porembski, R. Seine, and I. Theisen. 2007. *The Curious World of Carnivorous Plants: A Comprehensive Guide to Their Biology and Cultivation.* Timber Press: Portland, Oregon.

[40] "cocaine/crack" (http://www.urban75.com/Drugs/drugcoke.html). .

[41] "Deaths related to cocaine" (http://ar2005.emcdda.europa.eu/en/page050-en.html). .

[42] "Illegal drugs drain $160 billion a year from American economy" (http://web.archive.org/web/20080215071055/http://www. whitehousedrugpolicy.gov/NEWS/press02/012302.html). Archived from the original (http://www.whitehousedrugpolicy.gov/NEWS/ press02/012302.html) on 2008-02-15. .

[43] "The social cost of illegal drug consumption in Spain" (http://www.ingentaconnect.com/content/bsc/add/2002/00000097/00000009/ art00012). .

Further reading

General

- Evans, L. T. (1998). *Feeding the Ten Billion - Plants and Population Growth*. Cambridge University Press. Paperback, 247 pages. ISBN 0-521-64685-5.
- Kenrick, Paul & Crane, Peter R. (1997). *The Origin and Early Diversification of Land Plants: A Cladistic Study*. Washington, D. C.: Smithsonian Institution Press. ISBN 1-56098-730-8.
- Raven, Peter H., Evert, Ray F., & Eichhorn, Susan E. (2005). *Biology of Plants* (7th ed.). New York: W. H. Freeman and Company. ISBN 0-7167-1007-2.
- Taylor, Thomas N. & Taylor, Edith L. (1993). *The Biology and Evolution of Fossil Plants*. Englewood Cliffs, NJ: Prentice Hall. ISBN 0-13-651589-4.
- Trewavas A (2003). "Aspects of Plant Intelligence" (http://aob.oxfordjournals.org/cgi/content/full/92/1/1). *Annals of Botany* **92**: 1–20.

Species estimates and counts

- International Union for Conservation of Nature and Natural Resources (IUCN) Species Survival Commission (2004). IUCN Red List (http://www.iucnredlist.org/).
- Prance G. T. (2001). "Discovering the Plant World". *Taxon* **50**: 345–359.

External links

- *Plant* (http://www.eol.org/pages/281) at the Encyclopedia of Life
- Chaw, S.-M. et al. (1997). "Molecular Phylogeny of Extant Gymnosperms and Seed Plant Evolution: Analysis of Nuclear 18s rRNA Sequences" (http://mbe.library.arizona.edu/data/1997/1401/7chaw.pdf). *Molec. Biol. Evol.* **14** (1): 56–68. PMID 9000754.
- Index Nominum Algarum (http://ucjeps.berkeley.edu/INA.html)
- Interactive Cronquist classification (http://florabase.calm.wa.gov.au/phylogeny/cronq88.html)
- Plant Photo Gallery of Japan (http://www.alpine-plants-jp.com/art/index_photo2b.htm) - Flavon's Wild herb and Alpine plants
- Plant Picture Gallery (http://www.pflanzenliebe.de/)
- Plant Resources of Tropical Africa (http://www.prota.org/uk/About+Prota/)
- www.prota.org - PROTA's mission (http://database.prota.org/search.htm)
- Tree of Life (http://tolweb.org/Green_plants)

Botanical and vegetation databases

- African Plants Initiative database (http://www.aluka.org/action/doBrowse?sa=1&sa_sel=)
- Australia (http://www.anbg.gov.au/cpbr/databases/)
- Chilean plants at *Chilebosque* (http://www.chilebosque.cl/)
- Dave's garden (http://davesgarden.com/pf/) plenty of information mostly about garden plants
- e-Floras (Flora of China, Flora of North America and others) (http://www.efloras.org/index.aspx)
- Flora Europaea (http://rbg-web2.rbge.org.uk/FE/fe.html)
- Flora of Central Europe (http://www.floraweb.de/) (German)
- Flora of North America (http://www.efloras.org/flora_page.aspx?flora_id=1)
- List of Japanese Wild Plants Online (http://www.alpine-plants-jp.com/botanical_name/ list_of_japanese_wild_plants_abelia_buxus.htm)
- Meet the Plants-National Tropical Botanical Garden (http://www.ntbg.org/plants/choose_a_plant.php)
- Lady Bird Johnson Wildflower Center - Native Plant Information Network at University of Texas, Austin (http://www.wildflower.org/)
- The Plant List (http://www.theplantlist.org/)

- United States Department of Agriculture (http://plants.usda.gov/) not limited to continental US species

Biological_classification

Biological classification, or *scientific classification in biology*, is a method to group and categorize organisms into groups such as genus or species. These groups are known as **taxa** (singular: **taxon**). Biological classification is part of scientific taxonomy.

Modern biological classification has its root in the work of Carolus Linnaeus, who grouped species according to shared physical characteristics. These groupings have since been revised to improve consistency with the Darwinian principle of common descent. With the introduction of the cladistic method in the late 20th century, phylogenetic taxonomy in which organisms are grouped based purely on inferred evolutionary relatedness, ignoring morphological similarity, has become common in some areas of biology.[1] Molecular phylogenetics, which uses DNA sequences as data, has also driven many recent revisions and is likely to continue doing so. Biological classification belongs to the science of biological systematics.

Definition

Classification has been defined by Ernst Mayr as "The arrangement of entities in a hierarchical series of nested classes, in which similar or related classes at one hierarchical level are combined comprehensively into more inclusive classes at the next higher level." A class is defined as "a collection of similar entities".[2] (Note that the word "class" is used quite separately for one of the levels in the biological hierarchy.)

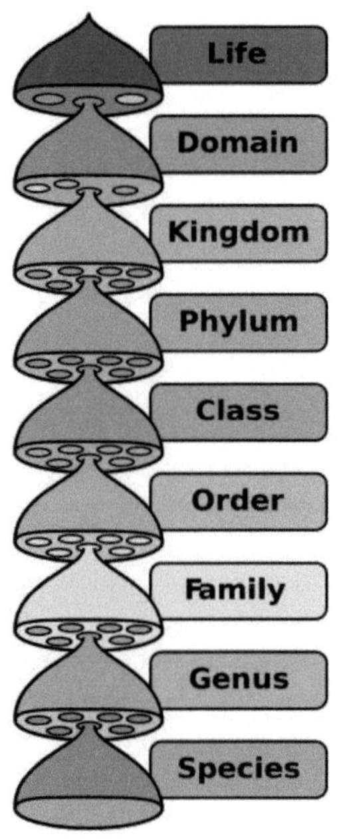

The hierarchy of biological classification's eight major taxonomic ranks, which is an example of definition by genus and differentia. Intermediate minor rankings are not shown.

What makes biological classification different from other classification systems (e.g. classifying books in a library) is evolution: the similarity between organisms placed in the same taxon is not arbitrary, but is instead a result of shared descent from their nearest common ancestor. Accordingly, the important attributes or traits for biological classification are 'homologous', i.e., inherited from common ancestors.[3] These must be separated from traits that are analogous. Thus birds and bats both have the power of flight, but this similarity is not used to classify them into a taxon (a "class"), because it is not inherited from a common ancestor. In spite of all the other differences between them, the fact that bats and whales both feed their young on milk is one of the features used to classify both of them as mammals, since it was inherited from a common ancestor(s).

Determining whether similarities are homologous or analogous can be difficult. Thus until recently, golden moles, found in South Africa, were placed in the same taxon (insectivores) as Northern Hemisphere moles, on the basis of morphological and behavioural similarities. However, molecular analysis has shown that they are not closely related, so that their similarities must be due to convergent evolution and not to shared descent, and so should not be used to place them in the same taxon.[4]

Biological types

The scientific names of taxa are formally attached to a **type**, which is one particular specimen (or in some cases a group of specimens, or in some cases an illustration) of the organism, preserved in a museum. The type is the example that serves to anchor or centralize the defining features of each particular taxon.

Taxonomic ranks

A classification as defined above is hierarchical. In a biological classification, **rank** is the level (the relative position) in a hierarchy. (Rarely, the term "taxonomic category" is used instead of "rank".) The *International Code of Zoological Nomenclature* defines rank, in the nomenclatural sense, as:

> The level, for nomenclatural purposes, of a taxon in a taxonomic hierarchy (e.g. all families are for nomenclatural purposes at the same rank, which lies between superfamily and subfamily).[5]

There are seven main ranks defined by the international nomenclature codes: kingdom, phylum/division, class, order, family, genus, species. "Domain", a level above kingdom, has become popular in recent years, but has not been accepted into the codes. Ranks between the seven main ones can be produced by adding prefixes such as "super-", "sub-" or "infra-". Thus a subclass has a rank between class and order, a superfamily between order and family. There are slightly different ranks for zoology and for botany, including subdivisions such as tribe.

Ranks are somewhat arbitrary, but hope to encapsulate the diversity contained within a group — a rough measure of the number of diversifications that the group has been through.[6]

Early systems

Ancient through medieval times

Current systems of classifying forms of life descend from the thought presented by the Greek philosopher Aristotle, who published in his metaphysical works the first known classification of everything whatsoever, or "being". This is the scheme that gave such words as 'substance', 'species' and 'genus' and was retained in modified and less general form by Linnaeus.

Aristotle also studied animals and classified them according to method of reproduction, as did Linnaeus later with plants. Aristotle's animal classification was eventually made obsolete by additional knowledge and forgotten.

The philosophical classification is in brief as follows:[7] Primary *substance* is the individual being; for example, Peter, Paul, etc. Secondary substance is a predicate that can properly or characteristically be said of a class of primary substances; for example, man of Peter, Paul, etc. The characteristic must not be merely in the individual; for example, being skilled in grammar.

Aristotle, 384–322 BC.

Grammatical skill leaves most of Peter out and therefore is not characteristic of him. Similarly man (all of mankind) is not in Peter; rather, he is in man.

Species is the secondary substance that is most proper to its individuals. The most characteristic thing that can be said of Peter is that Peter is a man. An identity is being postulated: "man" is equal to all its individuals and only those individuals. Members of a species differ only in number but are totally the same type.

Genus is a secondary substance less characteristic of and more general than the species; for example, man is an animal, but not all animals are men. It is clear that a genus contains species. There is no limit to the number of Aristotelian genera that might be found to contain the species. Aristotle does not structure the genera into phylum, class, etc., as the Linnaean classification does.

The secondary substance that distinguishes one species from another within a genus is the specific difference. Man can thus be comprehended as the sum of specific differences (the "differentiae" of biology) in less and less general categories. This sum is the definition; for example, man is an animate, sensate, rational substance. The most characteristic definition contains the species and the next most general genus: man is a rational animal. Definition is thus based on the unity problem: the species is but one yet has many differentiae.

The very top genera are the categories. There are ten: one of substance and nine of "accidents", universals that must be "in" a substance. Substances exist by themselves; accidents are only in them: quantity, quality, etc. There is no higher category, "being", because of the following problem, which was only solved in the Middle Ages by Thomas Aquinas: a specific difference is not characteristic of its genus. If man is a rational animal, then rationality is not a property of animals. Substance therefore cannot be a *kind* of being because it can have no specific difference, which would have to be *non*-being.

The problem of "being" occupied the attention of scholastics during the time of the Middle Ages. The solution of St. Thomas, termed the analogy of being, established the field of ontology, which received the better part of the publicity and also drew the line between philosophy and experimental science. The latter rose in the Renaissance from practical technique. Linnaeus, a classical scholar, combined the two on the threshold of the neo-classicist revival now called the Age of Enlightenment.

Renaissance through Age of Reason

An important advance was made by the Swiss professor, Conrad von Gesner (1516–1565). Gesner's work was a critical compilation of life known at the time.

The exploration of parts of the New World by Europeans produced large numbers of new plants and animals that needed descriptions and classification. The old systems made it difficult to study and locate all these new specimens within a collection and often the same plants or animals were given different names simply because there were too many species to keep track of. A system was needed that could group these specimens together so they could be found; the binomial system was developed based on morphology with groups having similar appearances. In the latter part of the

Rhinoceros in Conrad Gesner's *Historiae animalium*, 1551

16th century and the beginning of the 17th, careful study of animals commenced, which, directed first to familiar kinds, was gradually extended until it formed a sufficient body of knowledge to serve as an anatomical basis for classification. Advances in using this knowledge to classify living beings bear a debt to the research of medical anatomists, such as Fabricius (1537–1619), Petrus Severinus (1580–1656), William Harvey (1578–1657), and Edward Tyson (1649–1708). Advances in classification due to the work of entomologists and the first microscopists is due to the research of people like Marcello Malpighi (1628–1694), Jan Swammerdam (1637–1680), and Robert Hooke (1635–1702). Lord Monboddo (1714–1799) was one of the early abstract thinkers whose works illustrate knowledge of species relationships and who foreshadowed the theory of evolution.[8]

Early methodists

Since late in the 15th century, a number of authors had become concerned with what they called *methodus,* (method). By method authors mean an arrangement of minerals, plants, and animals according to the principles of logical division. The term *Methodists* was coined by Carolus Linnaeus in his *Bibliotheca Botanica* to denote the authors who care about the principles of classification (in contrast to the mere *collectors* who are concerned primarily with the description of plants paying little or no attention to their arrangement into genera, etc.). Important early Methodists were Italian philosopher, physician, and botanist Andrea Caesalpino, English naturalist John Ray, German physician and botanist Augustus Quirinus Rivinus, and French physician, botanist, and traveller Joseph Pitton de Tournefort.

Andrea Caesalpino (1519–1603) in his *De plantis libri XVI* (1583) proposed the first methodical arrangement of plants. On the basis of the structure of trunk and fructification he divided plants into fifteen "higher genera".

John Ray (1627–1705) was an English naturalist who published important works on plants, animals, and natural theology. The approach he took to the classification of plants in his Historia Plantarum was an important step towards modern taxonomy. Ray rejected the system of dichotomous division by which species were classified according to a pre-conceived, either/or type system, and instead classified plants according to similarities and differences that emerged from observation.

Both Caesalpino and Ray used traditional plant names and thus, the name of a plant did not reflect its taxonomic position (e.g. even though the apple and the peach belonged to different "higher genera" of John Ray's *methodus,* both retained their traditional names *Malus* and *Malus Persica* respectively). A further step was taken by Rivinus and Pitton de Tournefort who made genus a distinct rank within taxonomic hierarchy and introduced the practice of naming the plants according to their genera.

Augustus Quirinus Rivinus (1652–1723), in his classification of plants based on the characters of the flower, introduced the category of order (corresponding to the "higher" genera of John Ray and Andrea Caesalpino). He was the first to abolish the ancient division of plants into herbs and trees and insisted that the true method of division should be based on the parts of the fructification alone. Rivinus extensively used dichotomous keys to define both orders and genera. His method of naming plant species resembled that of Joseph Pitton de Tournefort. The names of all plants belonging to the same genus should begin with the same word (generic name). In the genera containing more than one species the first species was named with generic name only, while the second, etc. were named with a combination of the generic name and a modifier (*differentia specifica*).

Joseph Pitton de Tournefort (1656–1708) introduced an even more sophisticated hierarchy of class, section, genus, and species. He was the first to use consistently the uniformly composed species names that consisted of a generic name and a many-worded diagnostic phrase *differentia specifica.* Unlike Rivinus, he used *differentiae* with all species of polytypic genera.

Linnaean taxonomy

Carolus Linnaeus' great work, the *Systema Naturæ* (1st ed. 1735), ran through twelve editions during his lifetime. In this work, nature was divided into three kingdoms: mineral, vegetable and animal. Linnaeus used five ranks: class, order, genus, species, and variety.

Carolus Linnaeus

He abandoned long descriptive names of classes and orders still used by his immediate predecessors (Rivinus and Pitton de Tournefort) and replaced them with single-word names, provided genera with detailed diagnoses (*characteres naturales*), and combined numerous varieties into their species, thus saving botany from the chaos of new forms produced by horticulturalists.

Linnaeus is best known for his introduction of the method still used to formulate the scientific name of every species. Before Linnaeus, long many-worded names (composed of a generic name and a *differentia specifica*) had been used, but as these names gave a description of the species, they were not fixed. In his *Philosophia Botanica* (1751) Linnaeus took every effort to improve the composition and reduce the length of the many-worded names by abolishing unnecessary rhetorics, introducing new descriptive terms and defining their meaning with an unprecedented precision. In the late 1740s Linnaeus began to use a parallel system of naming species with *nomina trivialia*. *Nomen triviale*, a trivial name, was a single- or two-word epithet placed on the margin of the page next to the many-worded "scientific" name. The only rules Linnaeus applied to them was that the trivial names should be short, unique within a given genus, and that they should not be changed. Linnaeus consistently applied *nomina trivialia* to the species of plants in *Species Plantarum* (1st edn. 1753) and to the species of animals in the 10th edition of *Systema Naturæ* (1758).

By consistently using these specific epithets, Linnaeus separated nomenclature from description. Even though the parallel use of *nomina trivialia* and many-worded descriptive names continued until late in the eighteenth century, it was gradually replaced by the practice of using shorter proper names consisting of the generic name and the trivial name of the species. In the nineteenth century, this new practice was codified in the first Rules and Laws of Nomenclature, and the 1st edn. of *Species Plantarum* and the 10th edn. of *Systema Naturae* were chosen as starting points for the Botanical and Zoological Nomenclature respectively. This convention for naming species is referred to as binomial nomenclature.

Today, nomenclature is regulated by Nomenclature Codes, which allows names divided into taxonomic ranks.

Modern system

Whereas Linnaeus classified for ease of identification, the idea of the Linnaean taxonomy as translating into a sort of dendrogram of the Animal- and Plant Kingdoms was formulated toward the end of the 18th century, well before the *On the Origin of Species* was published. Among early works exploring the idea of a transmutation of species was Erasmus Darwin's 1796 Zoönomia and Jean-Baptiste Lamarck's Philosophie Zoologique of 1809. The idea was popularised in the Anglophone world by the speculative, but widely read Vestiges of the Natural History of Creation, published anonymously by Robert Chambers in 1844.[9]

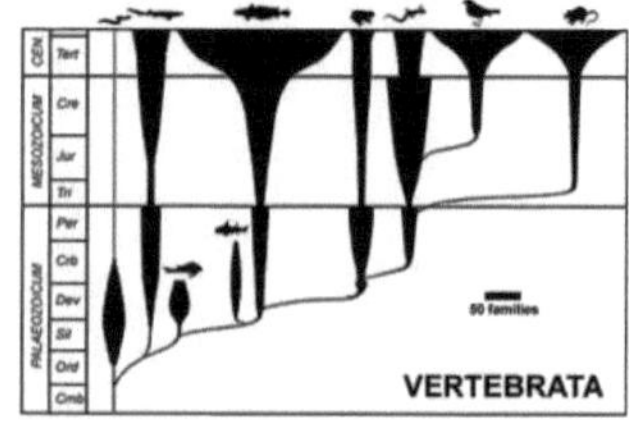

Evolution of the vertebrates at class level, width of spindles indicating number of families. Spindle diagrams are typical for Evolutionary taxonomy

With Darwin's theory, a general acceptance that classification should reflect the Darwinian principle of common descent quickly appeared. Tree of Life representations became popular in scientific works, with known fossil groups incorporated. One of the first modern groups tied to fossil ancestors were birds. Using the then newly discovered fossils of *Archaeopteryx* and *Hesperornis*, Thomas Henry Huxley pronounced that they had evolved from dinosaurs, a group formally named by Richard Owen in 1842.[10] The resulting description, that of dinosaurs

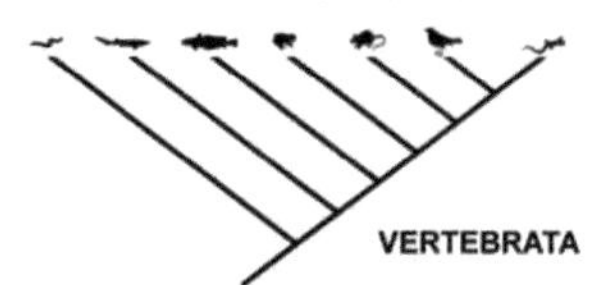

The same relationship, expressed as a cladogram typical for cladistics

"giving rise to" or being "the ancestors of" birds, is the essential hallmark of evolutionary taxonomic thinking. As more and more fossil groups were found and recognized in the late 19th and early 20th century, palaeontologists worked to understand the history of animals through the ages by linking together known groups[11] With the modern evolutionary synthesis of the early 1940s, an essentially modern understanding of evolution of the major groups was in place. The evolutionary taxonomy being based on Linnaean taxonomic ranks, the two terms are largely interchangeable in modern use.

Since the 1960s a trend called phylogenetic nomenclature (or cladism) has emerged, inspired by the cladistic method. The salient feature is arranging taxa in a hierarchical evolutionary tree, ignoring ranks. If a taxon includes all the descendants of some ancestral form, it is called monophyletic. Groups that have descendant groups removed from them (e.g. dinosaurs, with birds as offspring group) are termed paraphyletic, while groups representing more than one branch from the tree of life are called polyphyletic. A formal code of nomenclature, the *International Code of Phylogenetic Nomenclature*, or *PhyloCode* for short, is currently under development, intended to deal with names of clades. Linnaean ranks will be optional under the *PhyloCode*, which is intended to coexist with the current, rank-based codes.

Kingdoms and domains

From well before Linnaeus, plants and animals were considered separate Kingdoms. Linnaeus used this as the top rank, dividing the physical world into the plant, animal and mineral kingdoms. As advances in microscopy made classification of microorganisms possible, the number of kingdoms increased, five and six-kingdom systems being the most common.

Domains are a relatively new grouping. The three-domain system was first proposed in 1990, but not generally accepted until later. One main characteristic of the three-domain method is the separation of Archaea and Bacteria, previously grouped into the single kingdom Bacteria (a kingdom also sometimes called Monera). Consequently, the three domains of life are conceptualized as Archaea, Bacteria, and Eukaryota (comprising the nuclei-bearing eukaryotes).[12] A small minority of scientists add Archaea as a sixth kingdom, but do not accept the domain method.

Thomas Cavalier-Smith, who has published extensively on the classification of protists, has recently proposed that the Neomura, the clade that groups together the Archaea and Eukarya, would have evolved from Bacteria, more precisely from Actinobacteria. His classification of 2004 treats the archaebacteria as part of a subkingdom of the Kingdom Bacteria, i.e. he rejects the three-domain system entirely.[13]

Linnaeus 1735[14] 2 kingdoms	Haeckel 1866[15] 3 kingdoms	Chatton 1925[16] [17] 2 empires	Copeland 1938[18] [19] 4 kingdoms	Whittaker 1969[20] 5 kingdoms	Woese et al. 1977[21] [22] 6 kingdoms	Woese et al. 1990[23] 3 domains	Cavalier-Smith 2004[13] 6 kingdoms
(not treated)	Protista	Prokaryota	Monera	Monera	Eubacteria	Bacteria	Bacteria
					Archaebacteria	Archaea	
		Eukaryota	Protoctista	Protista	Protista	Eukarya	Protozoa
							Chromista
Vegetabilia	Plantae		Plantae	Plantae	Plantae		Plantae
			Protoctista	Fungi	Fungi		Fungi
Animalia	Animalia		Animalia	Animalia	Animalia		Animalia

Authorities (author citation)

An "authority" may be placed after a scientific name. The authority is the name of the scientist who first validly published the name. For example, in 1758 Linnaeus gave the Asian elephant the scientific name *Elephas maximus*, so the name is sometimes written as "*Elephas maximus* Linnaeus, 1758". The names of authors are frequently abbreviated: the abbreviation "L." is universally accepted for Linnaeus, and in botany there is a regulated list of standard abbreviations (see list of botanists by author abbreviation). The system for assigning authorities differs slightly between botany and zoology. However, it is standard that if a species' name or placement has been changed since the original description, the original authority's name is placed in parentheses.

Globally unique identifiers for names

There is a movement within the biodiversity informatics community to provide globally unique identifiers in the form of Life Science Identifiers (LSID) for all biological names. This would allow authors to cite names unambiguously in electronic media and reduce the significance of errors in the spelling of names or the abbreviation of authority names. Three large nomenclatural databases (referred to as nomenclators) have already begun this process, these are Index Fungorum, International Plant Names Index and ZooBank. Other databases, that publish taxonomic rather than nomenclatural data, have also started using LSIDs to identify **taxa**. The key example of this is Catalogue of Life. In the next step in integration, these taxonomic databases will include references to the nomenclatural databases using LSIDs.

See also

- All Species Foundation
- Holotype
- International Code of Nomenclature for algae, fungi, and plants
- International Code of Zoological Nomenclature
- List of Latin and Greek words commonly used in systematic names
- Phenetics
- Phylogenetic tree
- Species description
- Tree of Life Web Project
- Trinomial nomenclature
- Type (biology)

- Virus classification

References

[1] Laurin, M. (2010). "The subjective nature of Linnaean categories and its impact in evolutionary biology and biodiversity studies" (http://www.ctoz.nl/ctz/vol79/nr04/art01). *Contributions to Zoology* **79** (4). . Retrieved 21 March 2012.

[2] Mayr, Ernst & Bock, W.J. (2002). "Classifications and other ordering systems". *J. Zool. Syst. Evol. Research* **40** (4): 169–94. doi:10.1046/j.1439-0469.2002.00211.x.

[3] Mayr & Bock 2002, p. 178

[4] Mayr & Bock 2002, p. 178ff

[5] International Commission on Zoological Nomenclature (1999) *International Code of Zoological Nomenclature. Fourth Edition.* - International Trust for Zoological Nomenclature, XXIX + 306 pp.

[6] Error: Bad DOI specified!

[7] *Categories* Section 5 and *Metaphysics* Book 6, but the terms are used in many places throughout the writings of Aristotle.

[8] "Nomina Circumscribentia Insectorum" (http://www.insecta.bio.pu.ru). . Retrieved 2008-10-09.

[9] Secord, James A. (2000). *Victorian Sensation: The Extraordinary Publication, Reception, and Secret Authorship of Vestiges of the Natural History of Creation* (http://www.press.uchicago.edu/cgi-bin/hfs.cgi/00/14098.ctl). Chicago: University of Chicago Press. ISBN 978-0-226-74410-0. .

[10] Huxley, T.H. (1876): Lectures on Evolution. *New York Tribune*. Extra. no 36. In Collected Essays IV: pp 46-138 original text w/ figures (http://aleph0.clarku.edu/huxley/CE4/LecEvol.html)

[11] Rudwick, M. J. S. (1985). *The Meaning of Fossils: Episodes in the History of Palaeontology*. University of Chicago Press. p. 24. ISBN 0226731030

[12] See especially pp. 45, 78 and 555 of Joel Cracraft and Michael J. Donaghue, eds. (2004). *Assembling the Tree of Life*. Oxford, England: Oxford University Press.

[13] Cavalier-Smith, T. (2004), "Only six kingdoms of life" (http://www.cladocera.de/protozoa/cavalier-smith_2004_prs.pdf), *Proc. R. Soc. Lond. B* **271**: 1251–62, doi:10.1098/rspb.2004.2705, PMC 1691724, PMID 15306349, , retrieved 2010-04-29

[14] C. Linnaeus (1735). *Systemae Naturae, sive regna tria naturae, systematics proposita per classes, ordines, genera & species.*

[15] E. Haeckel (1866). *Generelle Morphologie der Organismen*. Reimer, Berlin.

[16] É. Chatton (1925). "*Pansporella perplexa*. Réflexions sur la biologie et la phylogénie des protozoaires". *Ann. Sci. Nat. Zool* **10-VII**: 1–84.

[17] É. Chatton (1937). *Titres et Travaux Scientifiques (1906–1937)*. Sette, Sottano, Italy.

[18] H. Copeland (1938). "The kingdoms of organisms". *Quarterly Review of Biology* **13**: 383–420. doi:10.1086/394568.

[19] H. F. Copeland (1956). *The Classification of Lower Organisms*. Palo Alto: Pacific Books.

[20] Whittaker RH (January 1969). "New concepts of kingdoms of organisms". *Science* **163** (3863): 150–60. doi:10.1126/science.163.3863.150. PMID 5762760.

[21] C. R. Woese, W. E. Balch, L. J. Magrum, G. E. Fox and R. S. Wolfe (August 1977). "An ancient divergence among the bacteria". *Journal of Molecular Evolution* **9** (4): 305–311. doi:10.1007/BF01796092. PMID 408502.

[22] Woese CR, Fox GE (November 1977). "Phylogenetic structure of the prokaryotic domain: the primary kingdoms". *Proc. Natl. Acad. Sci. U.S.A.* **74** (11): 5088–90. doi:10.1073/pnas.74.11.5088. PMC 432104. PMID 270744.

[23] Woese C, Kandler O, Wheelis M (1990). "Towards a natural system of organisms: proposal for the domains Archaea, Bacteria, and Eucarya." (http://www.pnas.org/cgi/reprint/87/12/4576). *Proc Natl Acad Sci U S A* **87** (12): 4576–9. Bibcode 1990PNAS...87.4576W. doi:10.1073/pnas.87.12.4576. PMC 54159. PMID 2112744. .

Bibliography

- Atran, S. (1990). *Cognitive foundations of natural history: towards an anthropology of science*. Cambridge, England: Cambridge University Press. xii+360 pages. ISBN 0521372933, 0521372933.

- Larson, J. L. (1971). *Reason and experience. The representation of Natural Order in the work of Carl von Linne*. Berkeley, California: University of California Press. VII+171 pages.

- Mayr, Ernst & Bock, W.J. (2002). "Classifications and other ordering systems". *J. Zool. Syst. Evol. Research* **40** (4): 169–94. doi:10.1046/j.1439-0469.2002.00211.x.

- Schuh, R. T. and A. V. Z. Brower. (2009). *Biological Systematics: principles and applications (2nd edn.)* Cornell University Press xiii+311 pages. ISBN 978-0-8014-4799-0

- Species 2000 & ITIS Catalogue of Life 2008 (http://www.catalogueoflife.org/annual-checklist/2008/browse_taxa.php)

- Stafleau, F. A. (1971). *Linnaeus and the Linnaeans. The spreading of their ideas in systematic botany, 1753–1789*. Utrecht: Oosthoek. xvi+386 pages.

Flowering_plant

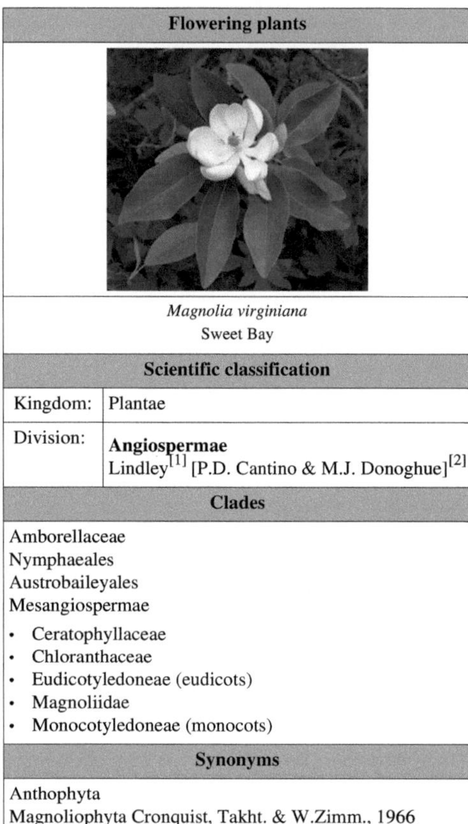

Flowering plants	
Magnolia virginiana Sweet Bay	
Scientific classification	
Kingdom:	Plantae
Division:	**Angiospermae** Lindley[1] [P.D. Cantino & M.J. Donoghue][2]
Clades	
Amborellaceae Nymphaeales Austrobaileyales Mesangiospermae • Ceratophyllaceae • Chloranthaceae • Eudicotyledoneae (eudicots) • Magnoliidae • Monocotyledoneae (monocots)	
Synonyms	
Anthophyta Magnoliophyta Cronquist, Takht. & W.Zimm., 1966	

The **flowering plants (angiosperms)**, also known as **Angiospermae** or **Magnoliophyta**, are the most diverse group of land plants. Angiosperms are seed-producing plants like the gymnosperms and can be distinguished from the gymnosperms by a series of synapomorphies (derived characteristics). These characteristics include flowers, endosperm within the seeds, and the production of fruits that contain the seeds.

The ancestors of flowering plants diverged from gymnosperms around 245–202 million years ago, and the first flowering plants known to exist are from 140 million years ago. They diversified enormously during the Lower Cretaceous and became widespread around 100 million years ago, but replaced conifers as the dominant trees only around 60–100 million years ago.

Angiosperm derived characteristics

Bud of a pink rose

- Flowers

The flowers, which are the reproductive organs of flowering plants, are the most remarkable feature distinguishing them from other seed plants. Flowers aid angiosperms by enabling a wider range of adaptability and broadening the ecological niches open to them. This has allowed flowering plants to largely dominate terrestrial ecosystems.

- Stamens with two pairs of pollen sacs

Stamens are much lighter than the corresponding organs of gymnosperms and have contributed to the diversification of angiosperms through time with adaptations to specialized pollination syndromes, such as particular pollinators. Stamens have also become modified through time to prevent self-fertilization, which has permitted further diversification, allowing angiosperms eventually to fill more niches.

- Reduced male parts, three cells

The male gametophyte in angiosperms is significantly reduced in size compared to those of gymnosperm seed plants. The smaller pollen decreases the time from pollination — the pollen grain reaching the female plant — to fertilization. In gymnosperms, fertilization can occur up to a year after pollination, whereas in angiosperms, fertilization begins very soon after pollination. The shorter time leads to angiosperm plants' setting seeds sooner and faster than gymnosperms, which is a distinct evolutionary advantage.

- Closed carpel enclosing the ovules (carpel or carpels and accessory parts may become the fruit)

The closed carpel of angiosperms also allows adaptations to specialized pollination syndromes and controls. This helps to prevent self-fertilization, thereby maintaining increased diversity. Once the ovary is fertilized, the carpel and some surrounding tissues develop into a fruit. This fruit often serves as an attractant to seed-dispersing animals. The resulting cooperative relationship presents another advantage to angiosperms in the process of dispersal.

- Reduced female gametophyte, seven cells with eight nuclei

The reduced female gametophyte, like the reduced male gametophyte, may be an adaptation allowing for more rapid seed set, eventually leading to such flowering plant adaptations as annual herbaceous life-cycles, allowing the flowering plants to fill even more niches.

- Endosperm

In general, endosperm formation begins after fertilization and before the first division of the zygote. Endosperm is a highly nutritive tissue that can provide food for the developing embryo, the cotyledons, and sometimes the seedling when it first appears.

These distinguishing characteristics taken together have made the angiosperms the most diverse and numerous land plants and the most commercially important group to humans. The major exception to the dominance of terrestrial ecosystems by flowering plants is the coniferous forest.

Evolution

Further information: Evolutionary history of plants − Flowers

Fossilized spores suggest that higher plants (embryophytes) have lived on the land for at least 475 million years.[3] Early land plants reproduced sexually with flagellated, swimming sperm, like the green algae from which they evolved. An adaptation to terrestrialization was the development of upright meiosporangia for dispersal by spores to new habitats. This feature is lacking in the descendants of their nearest algal relatives, the Charophycean green algae. A later terrestrial adaptation took place with retention of the delicate, avascular sexual stage, the gametophyte, within the tissues of the vascular sporophyte. This occurred by spore germination within sporangia rather than spore release, as in non-seed plants. A current example of how this might have happened can be seen in the precocious spore germination in *Sellaginella*, the spike-moss. The result for the ancestors of angiosperms was enclosing them in a case, the seed. The first seed bearing plants, like the ginkgo, and conifers (such as pines and firs), did not produce flowers. The pollen grains (males) of *Ginkgo* and cycads produce a pair of flagellated, mobile sperm cells that "swim" down the developing pollen tube to the female and her eggs.

Flowers of *Malus sylvestris* (crab apple)

The apparently sudden appearance of relatively modern flowers in the fossil record initially posed such a problem for the theory of evolution that it was called an "*abominable mystery*" by Charles Darwin.[4] However, the fossil record has considerably grown since the time of Darwin, and recently discovered angiosperm fossils such as *Archaefructus*, along with further discoveries of fossil gymnosperms, suggest how angiosperm characteristics may have been acquired in a series of steps. Several groups of extinct gymnosperms, in particular seed ferns, have been proposed as the ancestors of flowering plants, but there is no continuous fossil evidence showing exactly how flowers evolved. Some older fossils, such as the upper Triassic *Sanmiguelia*, have been suggested. Based on current evidence, some propose that the ancestors of the angiosperms diverged from an unknown group of gymnosperms during the late Triassic (245–202 million years ago). A close relationship between angiosperms and gnetophytes, proposed on the basis of morphological evidence, has more recently been disputed on the basis of molecular evidence that suggest gnetophytes are instead more closely related to other gymnosperms.

The evolution of seed plants and later angiosperms appears to be the result of two distinct rounds of whole genome duplication events.[5] These occurred in 319 [6] million years ago and 192 [7] million years ago respectively.

The earliest known macrofossil confidently identified as an angiosperm, *Archaefructus liaoningensis*, is dated to about 125 million years BP (the Cretaceous period),[8] while pollen considered to be of angiosperm origin takes the fossil record back to about 130 million years BP. However, one study has suggested that the early-middle Jurassic plant *Schmeissneria*, traditionally considered a type of ginkgo, may be the earliest known angiosperm, or at least a close relative.[9] In addition, circumstantial chemical evidence has been found for the existence of angiosperms as early as 250 million years ago. Oleanane, a secondary metabolite produced by many flowering plants, has been found in Permian deposits of that age together with fossils of gigantopterids.[10] [11] Gigantopterids are a group of extinct seed plants that share many morphological traits with flowering plants, although they are not known to have been flowering plants themselves.

Recent DNA analysis based on molecular systematics [12] [13] showed that *Amborella trichopoda*, found on the Pacific island of New Caledonia, belongs to a sister group of the other flowering plants, and morphological studies [14] suggest that it has features that may have been characteristic of the earliest flowering plants.

The orders Amborellales, Nymphaeales, and Austrobaileyales diverged as separate lineages from the remaining angiosperm clade at a very early stage in flowering plant evolution.[15]

The great angiosperm radiation, when a great diversity of angiosperms appears in the fossil record, occurred in the mid-Cretaceous (approximately 100 million years ago). However, a study in 2007 estimated that the division of the five most recent (the genus *Ceratophyllum*, the family Chloranthaceae, the eudicots, the magnoliids, and the monocots) of the eight main groups occurred around 140 million years ago.[16] By the late Cretaceous, angiosperms appear to have dominated environments formerly occupied by ferns and cycadophytes, but large canopy-forming trees replaced conifers as the dominant trees only close to the end of the Cretaceous 65 millions years ago or even later, at the beginning of the Tertiary.[17] The radiation of herbaceous angiosperm occurred much later.[18] Yet, many fossil plants recognizable as belonging to modern families (including beech, oak, maple, and magnolia) appeared already at late Cretaceous.

It is generally assumed that the function of flowers, from the start, was to involve mobile animals in their reproduction processes. That is, pollen can be scattered even if the flower is not brightly colored or oddly shaped in a way that attracts animals; however, by expending the energy required to create such traits, angiosperms can enlist the aid of animals and, thus, reproduce more efficiently.

Island genetics provides one proposed explanation for the sudden, fully developed appearance of flowering plants. Island genetics is believed to be a common source of speciation in general, especially when it comes to radical adaptations that seem to have required inferior transitional forms. Flowering plants may have evolved in an isolated setting like an island or island chain, where the plants bearing them were able to develop a highly specialized relationship with some specific animal (a wasp, for example). Such a relationship, with a hypothetical wasp carrying pollen from one plant to another

Two bees on a flower head of Creeping Thistle, *Cirsium arvense*

much the way fig wasps do today, could result in the development of a high degree of specialization in both the plant(s) and their partners. Note that the wasp example is not incidental; bees, which, it is postulated, evolved specifically due to mutualistic plant relationships, are descended from wasps.

Animals are also involved in the distribution of seeds. Fruit, which is formed by the enlargement of flower parts, is frequently a seed-dispersal tool that attracts animals to eat or otherwise disturb it, incidentally scattering the seeds it contains (see frugivory). While many such mutualistic relationships remain too fragile to survive competition and to spread widely, flowering proved to be an unusually effective means of reproduction, spreading (whatever its origin) to become the dominant form of land plant life.

Flower ontogeny uses a combination of genes normally responsible for forming new shoots.[19] The most primitive flowers are thought to have had a variable number of flower parts, often separate from (but in contact with) each other. The flowers would have tended to grow in a spiral pattern, to be bisexual (in plants, this means both male and female parts on the same flower), and to be dominated by the ovary (female part). As flowers grew more advanced, some variations developed parts fused together, with a much more specific number and design, and with either specific sexes per flower or plant, or at least "ovary-inferior".

Flower evolution continues to the present day; modern flowers have been so profoundly influenced by humans that some of them cannot be pollinated in nature. Many modern, domesticated flowers used to be simple weeds, which sprouted only when the ground was disturbed. Some of them tended to grow with human crops, perhaps already having symbiotic companion plant relationships with them, and the prettiest did not get plucked because of their beauty, developing a dependence upon and special adaptation to human affection.[20]

A few palaeontologists have also come up with a theory that flowering plants, or angiosperms, might have evolved because of dinosaurs; in other words, they believe that dinosaurs "created" flowers. One of the theory's biggest proponents is Robert T. Bakker. He theorizes that herbivorous dinosaurs, with their eating habits, forced plants to find new ways to develop new adaptations, in order to avoid predation by herbivores.

Classification

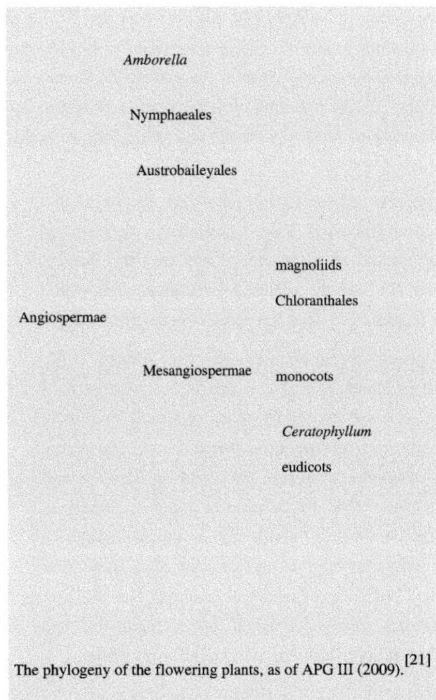

The phylogeny of the flowering plants, as of APG III (2009).[21]

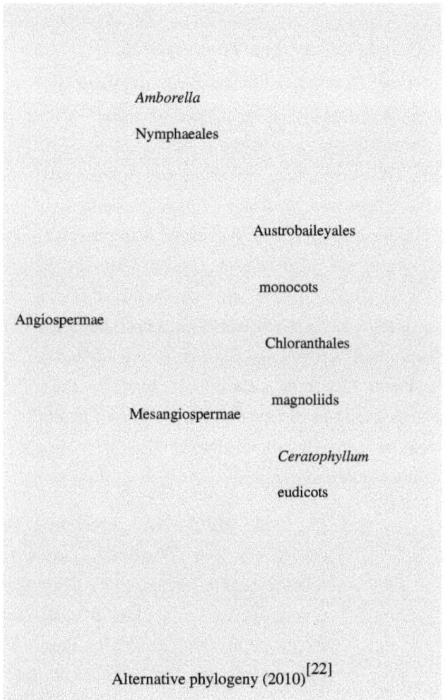

Alternative phylogeny (2010)[22]

There are eight groups of living angiosperms:

- *Amborella* — a single species of shrub from New Caledonia
- Nymphaeales — about 80 species[23] — water lilies and Hydatellaceae
- Austrobaileyales — about 100 species[23] of woody plants from various parts of the world
- Chloranthales — several dozen species of aromatic plants with toothed leaves
- Magnoliidae — about 9,000 species,[23] characterized by trimerous flowers, pollen with one pore, and usually branching-veined leaves — for example magnolias, bay laurel, and black pepper
- Monocotyledonae — about 70,000 species,[23] characterized by trimerous flowers, a single cotyledon, pollen with one pore, and usually parallel-veined leaves — for example grasses, orchids, and palms
- *Ceratophyllum* — about 6 species[23] of aquatic plants, perhaps most familiar as aquarium plants
- Eudicotyledonae — about 175,000 species,[23] characterized by 4- or 5- merous flowers, pollen with three pores, and usually branching-veined leaves — for example sunflowers, petunia, buttercup, apples and oaks

The exact relationship between these eight groups is not yet clear, although there is agreement that the first three groups to diverge from the ancestral angiosperm were Amborellales, Nymphaeales, and Austrobaileyales.[24] The term basal angiosperms refers to these three groups. The five other groups form the clade Mesangiospermae. The relationship between the three largest of these groups (magnoliids, monocots and eudicots) remains unclear. Some analyses make the magnoliids the first to diverge, others the monocots.[22] *Ceratophyllum* seems to group with the eudicots rather than with the monocots.

History of classification

The botanical term "Angiosperm", from the Ancient Greek αγγείον, *angeíon* (receptacle, vessel) and σπέρμα, (seed), was coined in the form Angiospermae by Paul Hermann in 1690, as the name of that one of his primary divisions of the plant kingdom. This included flowering plants possessing seeds enclosed in capsules, distinguished from his Gymnospermae, or flowering plants with achenial or schizo-carpic fruits, the whole fruit or each of its pieces being here regarded as a seed and naked. The term and its antonym were maintained by Carolus Linnaeus with the same sense, but with restricted application, in the names of the orders of his class Didynamia. Its use with any approach to its modern scope became possible only after 1827, when Robert Brown established the existence of truly naked ovules in the Cycadeae and Coniferae, and applied to them the name Gymnosperms. From that time onward, as long as these Gymnosperms were, as was usual, reckoned as dicotyledonous flowering plants, the term Angiosperm was used antithetically by botanical writers, with varying scope, as a group-name for other dicotyledonous plants.

From 1736, an illustration of Linnaean classification

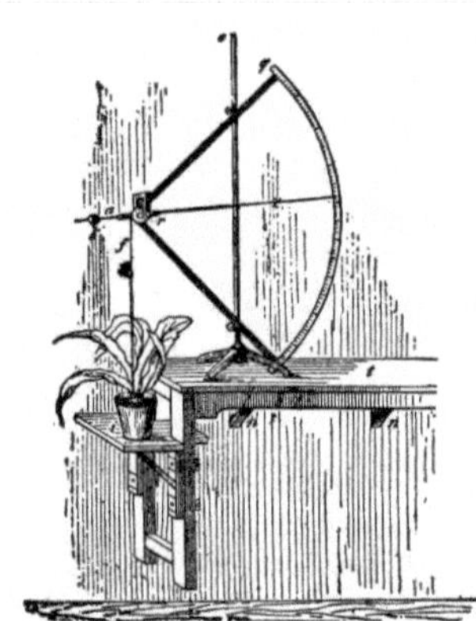

Auxanometer: Device for measuring increase or rate of growth in plants

In 1851, Hofmeister discovered the changes occurring in the embryo-sac of flowering plants, and determined the correct relationships of these to the Cryptogamia. This fixed the position of Gymnosperms as a class distinct from Dicotyledons, and the term Angiosperm then gradually came to be accepted as the suitable designation for the whole of the flowering plants other than Gymnosperms, including the classes of Dicotyledons and Monocotyledons. This is the sense in which the term is used today.

In most taxonomies, the flowering plants are treated as a coherent group. The most popular descriptive name has been Angiospermae (Angiosperms), with Anthophyta ("flowering plants") a second choice. These names are not linked to any rank. The Wettstein system and the Engler system use the name Angiospermae, at the assigned rank of subdivision. The Reveal system treated flowering plants as subdivision Magnoliophytina (Frohne & U. Jensen ex Reveal, Phytologia 79: 70 1996), but later split it to Magnoliopsida, Liliopsida, and Rosopsida. The Takhtajan system and Cronquist system treat this group at the rank of division, leading to the name Magnoliophyta (from the family name Magnoliaceae). The Dahlgren system and Thorne system (1992) treat this group at the rank of class, leading to the name Magnoliopsida. The APG system of 1998, and the later 2003[25] and 2009[21] revisions, treat the flowering plants as a clade called angiosperms without a formal botanical name. However, a formal classification was published alongside the 2009 revision in which the flowering plants form the Subclass Magnoliidae.[26]

The internal classification of this group has undergone considerable revision. The Cronquist system, proposed by Arthur Cronquist in 1968 and published in its full form in 1981, is still widely used but is no longer believed to accurately reflect phylogeny. A consensus about how the flowering plants should be arranged has recently begun to emerge through the work of the Angiosperm Phylogeny Group (APG), which published an influential reclassification of the angiosperms in 1998. Updates incorporating more recent research were published as APG II in 2003[25] and as APG III in 2009.[21] [27]

Traditionally, the flowering plants are divided into two groups, which in the Cronquist system are called *Magnoliopsida* (at the rank of class, formed from the family name *Magnoliacae*) and *Liliopsida* (at the rank of class, formed from the family name *Liliaceae*). Other descriptive names allowed by Article 16 of the ICBN include *Dicotyledones* or *Dicotyledoneae*, and *Monocotyledones* or *Monocotyledoneae*, which have a long history of use. In English a member of either group may be called a *dicotyledon* (plural *dicotyledons*) and *monocotyledon* (plural *monocotyledons*), or abbreviated, as *dicot* (plural *dicots*) and *monocot* (plural *monocots*). These names derive from the observation that the dicots most often have two *cotyledons*, or embryonic leaves, within each seed. The monocots usually have only one, but the rule is not absolute either way. From a diagnostic point of view, the number of cotyledons is neither a particularly handy nor a reliable character.

Monocot (left) and dicot seedlings

Recent studies, as by the APG, show that the monocots form a monophyletic group (clade) but that the dicots do not (they are paraphyletic). Nevertheless, the majority of dicot species do form a monophyletic group, called the *eudicots* or *tricolpates*. Of the remaining dicot species, most belong to a third major clade known as the Magnoliidae, containing about 9,000 species. The rest include a paraphyletic grouping of primitive species known collectively as the basal angiosperms, plus the families Ceratophyllaceae and Chloranthaceae.

Flowering plant diversity

The number of species of flowering plants is estimated to be in the range of 250,000 to 400,000.[28] [29] [30] The number of families in APG (1998) was 462. In APG II[25] (2003) it is not settled; at maximum it is 457, but within this number there are 55 optional segregates, so that the minimum number of families in this system is 402. In APG III (2009) there are 415 families.[21]

The diversity of flowering plants is not evenly distributed. Nearly all species belong to the eudicot (75%), monocot (23%) and magnoliid (2%) clades. The remaining 5 clades contain a little over 250 species in total, i.e., less than 0.1% of flowering plant diversity, divided among 9 families.

The most diverse families of flowering plants, in their APG circumscriptions, in order of number of species, are:

1. Asteraceae or Compositae (daisy family): 23,600 species[31]
2. Orchidaceae (orchid family): 22,075 species[31]
3. Fabaceae or Leguminosae (pea family): 19,400[31]
4. Rubiaceae (madder family): 13,150[32]
5. Poaceae or Gramineae (grass family): 10,035[31]
6. Lamiaceae or Labiatae (mint family): 7,173[31]
7. Euphorbiaceae (spurge family): 5,735[31]
8. Melastomataceae (melastome family): 5,005[31]
9. Myrtaceae (myrtle family): 4,620[31]
10. Apocynaceae (dogbane family): 4,555[31]

A poster of twelve different species of flowers of the *Asteraceae* family

In the list above (showing only the 10 largest families), the Orchidaceae and Poaceae are monocot families; the others are eudicot families.

Vascular anatomy

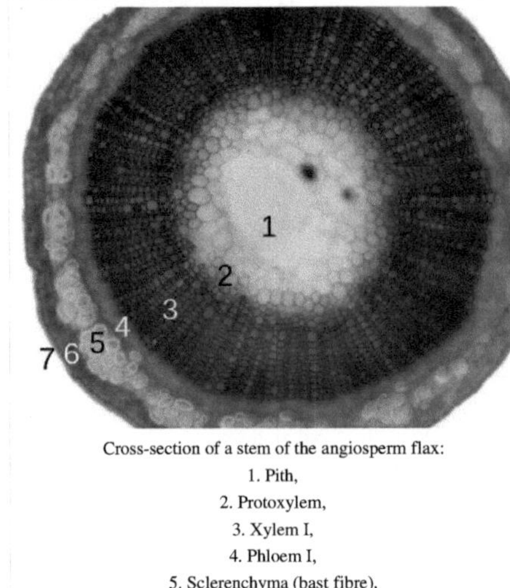

Cross-section of a stem of the angiosperm flax:
1. Pith,
2. Protoxylem,
3. Xylem I,
4. Phloem I,
5. Sclerenchyma (bast fibre),
6. Cortex,
7. Epidermis

The amount and complexity of tissue-formation in flowering plants exceeds that of gymnosperms. The vascular bundles of the stem are arranged such that the xylem and phloem form concentric rings.

In the dicotyledons, the bundles in the very young stem are arranged in an open ring, separating a central pith from an outer cortex. In each bundle, separating the xylem and phloem, is a layer of meristem or active formative tissue known as cambium. By the formation of a layer of cambium between the bundles (interfascicular cambium), a complete ring is formed, and a regular periodical increase in thickness results from the development of xylem on the inside and phloem on the outside. The soft phloem becomes crushed, but the hard wood persists and forms the bulk of the stem and branches of the woody perennial. Owing to differences in the character of the elements produced at the beginning and end of the season, the wood is marked out in transverse section into concentric rings, one for each season of growth, called annual rings.

Among the monocotyledons, the bundles are more numerous in the young stem and are scattered through the ground tissue. They contain no cambium and once formed the stem increases in diameter only in exceptional cases.

The flower, fruit, and seed

Flowers

The characteristic feature of angiosperms is the flower. Flowers show remarkable variation in form and elaboration, and provide the most trustworthy external characteristics for establishing relationships among angiosperm species. The function of the flower is to ensure fertilization of the ovule and development of fruit containing seeds. The floral apparatus may arise terminally on a shoot or from the axil of a leaf (where the petiole attaches to the stem). Occasionally, as in violets, a flower arises singly in the axil of an ordinary foliage-leaf. More typically, the flower-bearing portion of the plant is sharply distinguished from the foliage-bearing or vegetative portion, and forms a more or less elaborate branch-system called an inflorescence.

There are two kinds of reproductive cells produced by flowers. Microspores, which will divide to become pollen grains, are the "male" cells and are borne in the stamens (or microsporophylls). The "female" cells called megaspores, which will divide to become the egg cell (megagametogenesis), are contained in the ovule and enclosed in the carpel (or megasporophyll).

A collection of flowers forming an inflorescence

The flower may consist only of these parts, as in willow, where each flower comprises only a few stamens or two carpels. Usually, other structures are present and serve to protect the sporophylls and to form an envelope attractive to pollinators. The individual members of these surrounding structures are known as sepals and petals (or tepals in flowers such as *Magnolia* where sepals and petals are not distinguishable from each other). The outer series (calyx of sepals) is usually green and leaf-like, and functions to protect the rest of the flower, especially the bud. The inner series (corolla of petals) is, in general, white or brightly colored, and is more delicate in structure. It functions to attract insect or bird pollinators. Attraction is effected by color, scent, and nectar, which may be secreted in some part of the flower. The characteristics that attract pollinators account for the popularity of flowers and flowering plants among humans.

While the majority of flowers are perfect or hermaphrodite (having both pollen and ovule producing parts in the same flower structure), flowering plants have developed numerous morphological and physiological mechanisms to reduce or prevent self-fertilization. Heteromorphic flowers have short carpels and long stamens, or vice versa, so animal pollinators cannot easily transfer pollen to the pistil (receptive part of the carpel). Homomorphic flowers may employ a biochemical (physiological) mechanism called self-incompatibility to discriminate between self- and non-self pollen grains. In other species, the male and female parts are morphologically separated, developing on different flowers.

Fertilization and embryogenesis

Double fertilization refers to a process in which two sperm cells fertilize cells in the ovary. This process begins when a pollen grain adheres to the stigma of the pistil (female reproductive structure), germinates, and grows a long pollen tube. While this pollen tube is growing, a haploid generative cell travels down the tube behind the tube nucleus. The generative cell divides by mitosis to produce two haploid (n) sperm cells. As the pollen tube grows, it makes its way from the stigma, down the style and into the ovary. Here the pollen tube reaches the micropyle of the ovule and digests its way into one of the synergids, releasing its contents (which include the sperm cells). The synergid that the cells were released into degenerates and one sperm makes its way to fertilize the egg cell, producing a diploid ($2n$) zygote. The second sperm cell fuses with both central cell nuclei, producing a triploid ($3n$) cell. As the zygote develops into an embryo, the triploid cell develops into the endosperm, which serves as the embryo's food supply. The ovary now will develop into fruit and the ovule will develop into seed.

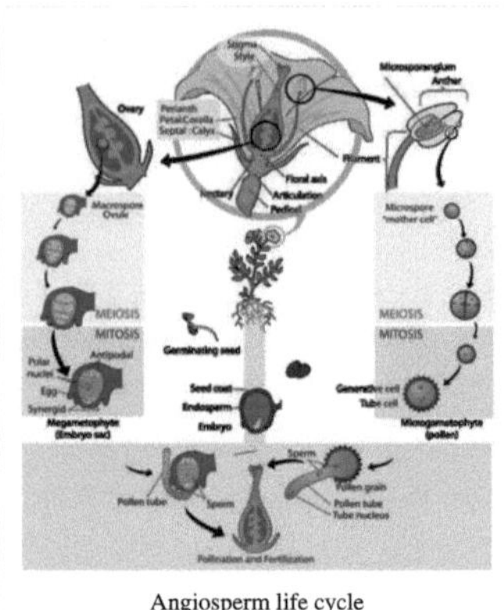

Angiosperm life cycle

Fruit and seed

The fruit of the *Aesculus* or Horse Chestnut tree

As the development of embryo and endosperm proceeds within the embryo sac, the sac wall enlarges and combines with the nucellus (which is likewise enlarging) and the integument to form the *seed coat*. The ovary wall develops to form the fruit or pericarp, whose form is closely associated with the manner of distribution of the seed.

Frequently, the influence of fertilization is felt beyond the ovary, and other parts of the flower take part in the formation of the fruit, e.g., the floral receptacle in the apple, strawberry, and others.

The character of the seed coat bears a definite relation to that of the fruit. They protect the embryo and aid in dissemination; they may also directly promote germination. Among plants with indehiscent fruits, in general, the fruit provides protection for the embryo and secures dissemination. In this case, the seed coat is only slightly developed. If the fruit is dehiscent and the seed is exposed, in general, the seed-coat is well developed, and must discharge the functions otherwise executed by the fruit.

Economic importance

Agriculture is almost entirely dependent upon angiosperms, which also provide a significant amount of livestock feed. Of all the families plants, the Poaceae, or grass family, is by far the most important, providing the bulk of all feedstocks (rice, corn — maize, wheat, barley, rye, oats, pearl millet, sugar cane, sorghum). The Fabaceae, or legume family, comes in second place. Also of high importance are the Solanaceae, or nightshade family (potatoes, tomatoes, and peppers, among others), the Cucurbitaceae, or gourd family (also including pumpkins and melons), the Brassicaceae, or mustard plant family (including rapeseed and the innumerable varieties of the cabbage species *Brassica oleracea*), and the Apiaceae, or parsley family. Many of our fruits come from the Rutaceae, or rue family (including oranges, lemons, grapefruits, etc.), and the Rosaceae, or rose family (including apples, pears, cherries, apricots, plums, etc.).

In some parts of the world, certain single species assume paramount importance because of their variety of uses, for example the coconut (*Cocos nucifera*) on Pacific atolls, and the olive (*Olea europaea*) in the Mediterranean region.

Flowering plants also provide economic resources in the form of wood, paper, fiber (cotton, flax, and hemp, among others), medicines (digitalis, camphor), decorative and landscaping plants, and many other uses. The main area in which they are surpassed by other plants is timber production.

See also

- List of garden plants
- List of plants by common name
- List of plant orders
- List of systems of plant taxonomy

References

[1] Lindley, J (1830). *Introduction to the Natural System of Botany*. London: Longman, Rees, Orme, Brown, and Green. xxxvi.

[2] Cantino, Philip D.; James A. Doyle, Sean W. Graham, Walter S. Judd, Richard G. Olmstead, Douglas E. Soltis, Pamela S. Soltis, & Michael J. Donoghue (2007). "Towards a phylogenetic nomenclature of *Tracheophyta*". *Taxon* **56** (3): E1–E44.

[3] Edwards, D (2000). "The role of mid-palaeozoic mesofossils in the detection of early bryophytes". *Philos Trans R Soc Lond B Biol Sci* **355** (1398): 733–755. doi:10.1098/rstb.2000.0613. PMC 1692787. PMID 10905607.

[4] Davies, T. J. (2004). "Darwin's abominable mystery: Insights from a supertree of the angiosperms". *Proceedings of the National Academy of Sciences* **101** (7): 1904–9. doi:10.1073/pnas.0308127100. PMC 357025. PMID 14766971.

[5] Jiao, Yuannian; Wickett, Norman J.; Ayyampalayam, Saravanaraj; Chanderbali, André S.; Landherr, Lena; Ralph, Paula E.; Tomsho, Lynn P.; Hu, Yi et al (2011). "Ancestral polyploidy in seed plants and angiosperms". *Nature* **473** (7345): 97–100. doi:10.1038/nature09916. PMID 21478875.

[6] http://toolserver.org/~verisimilus/Timeline/Timeline.php?Ma=319

[7] http://toolserver.org/~verisimilus/Timeline/Timeline.php?Ma=192

[8] Sun G., Ji Q., Dilcher D.L., Zheng S., Nixon K.C., Wang X. (2002). "Archaefructaceae, a New Basal Angiosperm Family" (http://www. sciencemag.org/cgi/content/abstract/296/5569/899?ck=nck&siteid=sci&ijkey=8dZ6zTqF606ps&keytype=ref). *Science* **296** (5569): 899–904. doi:10.1126/science.1069439. PMID 11988572. .

[9] Xin Wing; Shuying Duan, Baoyin Geng, Jinzhong Cui and Yong Yang (2007). "Schmeissneria: A missing link to angiosperms?". *BMC Evolutionary Biology* **7**: 14. doi:10.1186/1471-2148-7-14. PMC 1805421. PMID 17284326.

[10] Taylor, David Winship; Li, Hongqi; Dahl, Jeremy; Fago, Fred J.; Zinniker, David; Moldowan, J. Michael (2006). "Biogeochemical evidence for the presence of the angiosperm molecular fossil oleanane in Paleozoic and Mesozoic non-angiospermous fossils". *Paleobiology* **32** (2): 179. doi:10.1666/0094-8373(2006)32[179:BEFTPO]2.0.CO;2. ISSN 0094-8373.

[11] Oily Fossils Provide Clues To The Evolution Of Flowers (http://www.sciencedaily.com/releases/2001/04/010403071438.htm) — ScienceDaily (Apr. 5, 2001)

[12] NOVA — Transcripts — First Flower (http://www.pbs.org/wgbh/nova/transcripts/3405_flower.html) — PBS Airdate: April 17, 2007

[13] Soltis, D. E.; Soltis, P. S. (2004). "*Amborella* not a "basal angiosperm"? Not so fast". *American Journal of Botany* **91** (6): 997–1001. doi:10.3732/ajb.91.6.997. PMID 21653455.

[14] South Pacific plant may be missing link in evolution of flowering plants (http://www.eurekalert.org/pub_releases/2006-05/ uoca-spp051506.php) — Public release date: 17-May-2006

[15] Vialette-Guiraud, AC; Alaux, M; Legeai, F; Finet, C; Chambrier, P; Brown, SC; Chauvet, A; Magdalena, C et al (2011). "Cabomba as a model for studies of early angiosperm evolution". *Annals of botany* **108** (4): 589–98. doi:10.1093/aob/mcr088. PMC 3170152. PMID 21486926.

[16] Moore, M. J.; Bell, C. D.; Soltis, P. S.; Soltis, D. E. (2007). "Using plastid genome-scale data to resolve enigmatic relationships among basal angiosperms". *Proceedings of the National Academy of Sciences* **104** (49): 19363–8. doi:10.1073/pnas.0708072104. PMC 2148295. PMID 18048334.

[17] David Sadava; H. Craig Heller; Gordon H. Orians; William K. Purves, David M. Hillis (December 2006). *Life: the science of biology* (http:/ /books.google.com/books?id=1m0_FLEjd-cC&pg=PA477). Macmillan. pp. 477–. ISBN 9780716776741. . Retrieved 4 August 2010.

[18] Stewart, Wilson Nichols; Rothwell, Gar W. (1993). *Paleobotany and the evolution of plants* (2nd ed.). Cambridge Univ. Press. p. 498. ISBN 0521233151.

[19] Age-Old Question On Evolution Of Flowers Answered (http://unisci.com/stories/20012/0615015.htm) — 15-Jun-2001

[20] Human Affection Altered Evolution of Flowers (http://www.livescience.com/othernews/050526_flower_power.html) — By Robert Roy Britt, LiveScience Senior Writer (posted: 26 May 2005 06:53 am ET)

[21] Angiosperm Phylogeny Group (2009). "An update of the Angiosperm Phylogeny Group classification for the orders and families of
       flowering plants: APG III" (http://www3.interscience.wiley.com/journal/122630309/abstract). *Botanical Journal of the Linnean Society*
       161 (2): 105–121. doi:10.1111/j.1095-8339.2009.00996.x. . Retrieved 2010–12–10.

[22] Bell, C.D.; Soltis, D.E. & Soltis, P.S. (2010). "The Age and Diversification of the Angiosperms Revisited". *American Journal of Botany* **97**
       (8): 1296–1303. doi:10.3732/ajb.0900346. PMID 21616882., p. 1300

[23] Jeffrey D. Palmer, Douglas E. Soltis and Mark W. Chase, Chase, M. W. (2004). *Figure 2* (http://www.amjbot.org/cgi/content/full/91/
       10/1437/F2). "The plant tree of life: an overview and some points of view" (http://www.amjbot.org/cgi/content/full/91/10/1437).
       American Journal of Botany **91** (10): 1437–1445. doi:10.3732/ajb.91.10.1437. PMID 21652302. .

[24] Pamela S. Soltis and Douglas E. Soltis (2004). "The origin and diversification of angiosperms" (http://www.amjbot.org/cgi/content/full/
       91/10/1614). *American Journal of Botany* **91** (10): 1614–1626. doi:10.3732/ajb.91.10.1614. PMID 21652312. .

[25] Angiosperm Phylogeny Group (2003). "An update of the Angiosperm Phylogeny Group classification for the orders and families of
       flowering plants: APG II" (http://www.blackwell-synergy.com/links/doi/10.1046/j.1095-8339.2003.t01-1-00158.x/full/). *Botanical
       Journal of the Linnean Society* **141** (4): 399–436. doi:10.1046/j.1095-8339.2003.t01-1-00158.x. .

[26] Chase, Mark W. & Reveal, James L. (2009). "A phylogenetic classification of the land plants to accompany APG III". *Botanical Journal of
       the Linnean Society* **161** (2): 122–127. doi:10.1111/j.1095-8339.2009.01002.x. I Chase & Reveal 2009

[27] *As easy as APG III - Scientists revise the system of classifying flowering plants* (http://www.linnean.org/index.php?id=448). The Linnean
       Society of London. 2009-10-08. . Retrieved 2009-10-29.

[28] Thorne, R. F. (2002). "How many species of seed plants are there?" (http://www.ingentaconnect.com/content//iapt/tax/2002/
       00000051/00000003/art00009). *Taxon* **51** (3): 511–522. doi:10.2307/1554864. JSTOR 1554864. .>

[29] Scotland, R. W. & Wortley, A. H. (2003). "How many species of seed plants are there?" (http://www.ingentaconnect.com/content/iapt/
       tax/2003/00000052/00000001/art00011). *Taxon* **52** (1): 101–104. doi:10.2307/3647306. JSTOR 3647306. .

[30] Govaerts, R.url=http://www.ingentaconnect.com/content/iapt/tax/2003/00000052/00000003/art00016 (2003). "How
       many species of seed plants are there? — a response". *Taxon* **52** (3): 583–584. doi:10.2307/3647457. JSTOR 3647457.

[31] Stevens, P.F. (2001 onwards). "Angiosperm Phylogeny Website (at Missouri Botanical Garden)" (http://www.mobot.org/MOBOT/
       Research/APweb/welcome.html). .

[32] "Kew Scientist 30 (October2006)" (http://www.kew.org/kewscientist/ks_30.pdf). .

Further reading

- Cronquist, Arthur (1981). *An Integrated System of Classification of Flowering Plants*. New York: Columbia
  Univ. Press. ISBN 0-23-103880-1.
- Heywood, V. H., Brummitt, R. K., Culham, A. & Seberg, O. (2007). *Flowering Plant Families of the World*.
  Richmond Hill, Ontario, Canada: Firefly Books. ISBN 1-55407-206-9.
- Dilcher, D. (2000). "Toward a new synthesis: Major evolutionary trends in the angiosperm fossil record".
  Proceedings of the National Academy of Sciences **97** (13): 7030. doi:10.1073/pnas.97.13.7030.
- Simpson, M.G. *Plant Systematics*, 2nd Edition. Elsevier/Academic Press. 2010.
- Raven, P.H., R.F. Evert, S.E. Eichhorn. *Biology of Plants*, 7th Edition. W.H. Freeman. 2004.

External links

- Cole, Theodor C.H.; Hilger, Dr. Harmut H. Angiosperm Phylogeny Poster – Flowering Plant Systamatics (http://
  www.biologie.fu-berlin.de/sysbot/poster/poster1.pdf)
- Cromie, William J. (December 16, 1999). "Oldest Known Flowering Plants Identified By Genes" (http://www.
  news.harvard.edu/gazette/1999/12.16/angiosperms.html). Harvard University Gazette.
- Watson, L. and Dallwitz, M.J. (1992 onwards). The families of flowering plants: descriptions, illustrations,
  identification, information retrieval. (http://www.biologie.uni-hamburg.de/b-online/delta/angio/)
- *Flowering plant* (http://www.eol.org/pages/282) at the Encyclopedia of Life

Article Sources and Contributors

Sprengelia *Source*: http://en.wikipedia.org/w/index.php?title=Sprengelia *Contributors*: Berichard, Dysmorodrepanis, Melburnian, R'n'B

Ericaceae *Source*: http://en.wikipedia.org/w/index.php?title=Ericaceae *Contributors*: (, Algirdas, Atubeileh, BC Myles, BD2412, Ben-Zin, Bff, Bomac, BoundaryRider, CanisRufus, Chester Markel, Circeus, Codiferous, Conversion script, DanielCD, Derster, Dieter Simon, Dj Capricorn, Dyanega, Dysmorodrepanis, ENeville, EncycloPetey, Eug, FatBear1, Franz Xaver, Ganzy, Grahamec, Gruzd, Heidijane, Hesperian, Hr oskar, Imc, Island, JJ Harrison, Jagdfeld, Jaknouse, JoJan, Karen Johnson, Kazvorpal, Kingdon, Kiwikibble, Lavateraguy, Looxix, Loren Rosen, MPF, Marshman, Melburnian, Menchi, MidgleyDJ, Mike Dallwitz, Moflg, Mukogodo, My Cat inn, Nk, Nonenmac, Paul venter, PierreAbbat, Pion, PuzzletChung, Razimantv, Renato Caniatti, Reporein, Ricardo Carneiro Pires, Rich Farmbrough, Richard New Forest, Rjwilmsi, SchuminWeb, Seglea, Smartse, Sortior, Stan Shebs, Stemonitis, Tgeairn, Unyoyega, UtherSRG, Vuong Ngan Ha, WormRunner, Wsiegmund, , 22 anonymous edits

Ericales *Source*: http://en.wikipedia.org/w/index.php?title=Ericales *Contributors*: Adyione, AstroNomer, Atubeileh, Berton, Billfred, CanisRufus, Chase me ladies, I'm the Cavalry, Conversion script, DWeir, DanielCD, Eclecticology, Fanghong, Gabbe, Ginkgo100, Glenn, Gruzd, Hesperian, Iorsh, Ironholds, Jannex, Jmgarg1, JoJan, Josh Grosse, Karen Johnson, Kembangraps, Kingdon, Kurykh, Ligia, Ligulem-s, Looxix, Lord Voldemort, MPF, Martpol, Melaen, Naddy, Nathan R. Noll, Nicklott, Nk, Nono64, Peter G Werner, Quinxorin, Rjwilmsi, Sanfranman59, Stan Shebs, Stemonitis, TDogg310, Template namespace initialisation script, TeunSpaans, Tom Radulovich, UtherSRG, Vuong Ngan Ha, Wanderer099, WikHead, WooteleF, Yath, Zumost, يسرا, 25 anonymous edits

Asterids *Source*: http://en.wikipedia.org/w/index.php?title=Asterids *Contributors*: Bjankuloski06en, Borgx, Brya, ENeville, Fanghong, Floscuculi, Gvnn, Human anatomy, JNW, Johannes Lundberg, KOMPAN, Kleopatra, L Kensington, Lavateraguy, Lenticel, Logan, MrDarwin, N2e, Newone, Pinethicket, R. S. Shaw, Raz1el, Rkitko, Rosarinagazo, Vuong Ngan Ha, 10 anonymous edits

Eudicots *Source*: http://en.wikipedia.org/w/index.php?title=Eudicots *Contributors*: AslForwe, Betacommand, Billy13579Joe2, Bjankuloski06en, Brya, CRICKETLtd, CalicoCatLover, Cephal-odd, Csparr, CyrilThePig4, Dejvid, DrMicro, Dysmorodrepanis, Fanghong, Flammifer, Hohum, Hqb, Human anatomy, IOLJeff, Igoldste, JoJan, Josh Grosse, Kingdon, Kupirijo, Lavateraguy, Logan, MPF, Middayexpress, MrDarwin, Newone, NoahElhardt, Pomeapplepome, Psyche825, Qualia2, Raghith, Raz1el, Rkitko, SB Johnny, Snek01, Stemonitis, Syp, T34, Ucucha, Vuong Ngan Ha, Whilom, William Avery, Чръный человек, , 25 anonymous edits

Plant *Source*: http://en.wikipedia.org/w/index.php?title=Plant *Contributors*: 1nic, 321mister, Abductive, Acalamari, Accurizer, Ace ETP, Adam McMaster, Adambro, Adashiel, Adenosine, Adsomvilay, Agong1, Agrihouse, Ahoerstemeier, Alan Liefting, Aleksa Lukic, Ali, Alink, Almafeta, Almog6564, Altenmann, Alwolff55, AmiDaniel, Anclation, Andre Engels, Andrew R. White, Andrewjlockley, Andrewpmk, Andris, Androstachys, Andux, Andyjsmith, Anger22, Angielaj, Anna Frodesiak, Anomalocaris, Antandrus, AnthonyBryars, AntiVan, Antonrojo, Anwar saadat, AquamarineOnion, Arcadian, Arfan, Arjun01, Arjuna316, Arkuat, Arnorian, Arsalan daudi, Art LaPella, Ashley Y, Ashmoo, Astropithicus, AtomicDragon, AuburnPilot, Aude, Autocracy, AxelBoldt, AxiomShell, Banes, BanyanTree, Barbara Shack, Barklund, Barneca, Barticus88, BaseballDetective, Bdiscoe, Bdwolverine87, Bearly541, Bearman4, Beezhive, Benbest, Bendzh, Bensaccount, Bergsten, BerneyBoy, Bewareofdog, Bibliomaniac15, Billare, Bleach926, Bobianite, Bobisbob, Bobo192, Bomac, Bongwarrior, BorgQueen, BorisTM, Borislav, Bosniak, Brainster 101, Branddobbe, Brastein, Briséis, Bubba hotep, C.Fred, CIreland, Caca, Cadiomals, Caltas, Calvin Limuel, Can't sleep, clown will eat me, CanDo, CapitalR, Cephal-odd, Chefyingi, Chickenflicker, Chipmunkdavis, Chizeng, Christian List, Ciaoriki, Circeus, Ckatz, Clemwang, Cmdrjameson, Conscious, Conversion script, Cornflake4, Corp1117, Cratylus3, Crazychinchilla, Cremepuff222, Croat Canuck, Crusadeonilliteracy, Csparr, Ctbolt, Curtis Clark, CurtisLambert, Cwmhiraeth, Czj, DVD R W, Dacoutts, Daderot, Damieng, Dan East, Dan Guan, Dangerousnerd, DanielCD, Dannown, Dappawit, Darkwind, Db099221, Dcflyer, Dcoetzee, DeadEyeArrow, DeliciousT, Demmy, Deor, DerHexer, Dfrg.msc, Dinoguy2, Dipics, Directoryguru, Discospinster, Diucón, Dlloyd, Dlohcierekim, Doc glasgow, DocWatson42, Doloco, Dominus, Donald Albury, Donarreiskoffer, DougEDoug, Doyley, Dr.Bastedo, Dubaduba, Dv82matt, ESkog, Ed g2s, Edcolins, Edward321, Eleassar777, Eminem95941, Emo childnc1993, Emperorbma, EncycloPetey, EricG, Ettrig, Evercat, Everyking, F 22, FF2010, Fairsing, Farrahi, Femto, Filemon, Finlux, Fir0002, First Light, Floaterfluss, Fox6453, Franciepants18, FreeKresge, FreplySpang, Frogking51, Frymaster, Fumple3222, Funandtrvl, Fvasconcellos, Gabbe, Gaius Cornelius, Galoubet, Gdarin, Gdr, Geniac, Geo987, Geologyguy, Ghamer01, Giftlite, Gilliam, Gimmie, Giraffedata, Glenn, Gmunder, Gnangarra, Goatasaur, Gogo Dodo, Gop 24fan, Gotmilk954, Gracenotes, Graham87, Graminophile, Greatestrowerever, Guanaco, Guettarda, Gurch, Gveret Tered, Gwynplaine, Hadal, Haemo, Hall Monitor, Hallenrm, Halsteadk, Hapsiainen, Hardyplants, Hattes, Hdt83, Hdurina, Heegoop, Heimstern, Helikophis, HelloMojo, HenkvD, Henriette, HenryLi, Hephaestos, Heron, Heyitschrisss, Hi112233, Hi332211, Hmains, Hordaland, Hotty 8p, Hu12, Husond, Hyark, II MusLiM HyBRiD II, Ichormosquito, Ikanreed, Iloveplants, Infrogmation, Inner Earth, InvisibleK, IronGargoyle, Ischa1, Ixfd64, JForget, JK23, JLaTondre, JYi, Jackl, Jaknouse, Jamymc, Janetleopold, Jannex, Japs eye, Jason Leach, Jay2332, JayW, Jclerman, Jeepday, JemGage, Jemsherlock, JeremyA, Jergen, Jeronimo, Jerry, JesseGarrett, Jhagadurn, Jiddisch, JinJian, Jkyser, JoJan, Joan-of-arc, JoanneB, JoeSmack, Jolt76, Josh Grosse, Joyous!, Jpatros, Junglecat, Jurj, Juzeris, Jwanders, Jyril, Kakorot84, Karl-Henner, Karlthegreat, Karukera, Kdammers, Kesac, Ketchupabi, Kholiana joy, Kidpoker15, Kilva, Kim Bruning, Kingdon, Klemen Kocjancic, Kleopatra, Kniveswood, Kopro002, Kostisl, Kostya, Kowey, Kprwiki, Krinkle, Kungfuadam, Kuohatti, Kupirijo, Kuru, Kyleistwo, Lahiru k, Lancevortex, Lankiveil, LarryMorseDCOhio, Latitude0116, Lavateraguy, Lazulilasher, Lcgarcia, Le chien manquee, Leafyplant, Lenticel, Lesfreck, Lexor, Liface, Lights, Ligulem-s, Livestock kills, Llull, Lola lafanda, Looxix, Lord Emsworth, Lucyin, Luna Santin, Lunz2121, MER-C, MK8, MPF, MacGyverMagic, Magister Mathematicae, Magnus Manske, Maitch, Majorly, Malcolm Farmer, Man vyi, Marietta Georgia, Marrovi, MarsRover, Marshall111, Marshman, Matt Crypto, Matthew Wilson, Mav, Maxim, Mbz1, McMondongo, Mediamute, Melburnian, Menchi, Methcub, Mfatic, Mgiganteus1, Michaplot, Mike Rosoft, Million Moments, Minna Sora no Shita, Miss Madeline, Misza13, Moe Epsilon, Mofomojo, Moman, Monedula, Money4nothing, Monobi, Monolith224, Moon&Nature, Motorbikematt, Moyogo, Mr Chuckles, Mr Stephen, MrDarwin, MrFish, Msikma, Mxn, Mário e Dário, Naddy, Narayanese, Naryanlal, Natalie Erin, Natski23, NatureA16, Naveen Sankar, NawlinWiki, Nayvik, Neale Monks, Nerval, Netoholic, Neverquick, NewEnglandYankee, Nihuop, Ninjaboi887, Nivix, Njrfrog, Nk, Nohat, NorCal764, Northamerica1000, Nwbeeson, OMGrace, OOJaxxOo, Obli, Oblivion667, Ocatecir, Oceans and oceans, Ojaswi joshi, Oleg Alexandrov, Olivierd, Olyashok, Omicronpersei8, Onco p53, Op. Deo, Opelio, Optakeover, Osborne, Ospalh, Ost316, OwenX, Oxymoron83, P30Carl, PARA19, PKM, Paaerduag, Part Deux, Patstuart, Paul Murray, Paulbob, Pauli133, Pavel Vozenilek, Pb30, Pedroalexandrade, Pekaje, Per Ardua, Perfect Proposal, Persian Poet Gal, Peruvianllama, PetTrees, Peter G Werner, Peter coxhead, Petowl, Phaedriel, Phatvet21, Philip Trueman, PierreAbbat, Pippu d'Angelo, Plantguy, Plumbago, Poindexter Propellerhead, Polyhedron, Poopisme, Poor Yorick, Posem4012, Possum, Prodego, Pseudomonas, Pthag, PuzzletChung, Pyrospirit, Python eggs, QuartierLatin1968, R0uge, RadiantRay, RadicalBender, Radiojon, RaffiKojian, Raichu, Rainbowzorn, Rajivvkr, Randomuselesstrivia, Raymond arritt, Raz1el, Rdsmith4, RedRollerskate, Redattack34, Redvers, Reitzaleah, Rejax, Renegadeshark, Retiono Virginian, RexNL, Rhetoricofdos, Rhobite, Rich Farmbrough, Richard001, Rickproser, Righthonorablegentleman, Rjwilmsi, Rkitko, Rocastelo, Rockvee, Rommexidas97826, Rsamahamed, Rtkat3, Ryan-D, SASHHI, SB Johnny, SFODLONGBOW, SJP, SU Linguist, SWAdair, Saaga, Sackosacko, Sal817, Sam Hocevar, Samsara, Samuel, Samw, Sander Säde, Sander123, Sannse, Sarpasarpeti101, Savant13, Sceptre, Schzmo, Sci-Fi Dude, Scilit, Scohoust, Scottalter, Scoutersig, Seba, Sengkang, Setu, Sfmammamia, Shadowjams, Sherool, Shyamal, SilkTork, Silver Edge, Sinn, SirLeopold, Sjb1234567, Sjc, Sjwk, Sluzzelin, Smith609, Smooth O, Solipsist, Solitude, Somejan, Soumyasch, Spartan-James, SpeedyGonsales, SpookyMulder, Srose, Stedd, Stephenb, Stevertigo, Stewartadcock, Stizz, Styrofoam1994, Sugarsugar30, Sumergocognito, Summer Song, SuperTyphoon, Superruss, Sushant gupta, SuzanneKn, Sylverfysh, TUF-KAT, Tanaats, Tarret, Tau'olunga, Tawker, Tbhotch, TedE, Telecineguy, Telstra Robs, Tesseract2, The Anome, The Evil Spartan, The Genesis, The Mysterious El Willstro, The Random Editor, The elephantman, The undertow, TheAlphaWolf01, Theda, Theo148, Theresa knott, Thrallhascome, Tigershrike, TimVickers, Timir2, Timmy94, Titoxd, Tnoble, Tom Radulovich, Tom2005g, Torahjerus14, Tpbradbury, TreasuryTag, Trilobitealive, Tropylium, Turlo Lomon, Twoars, Ucanlookitup, Unleo, Unyoyega, Usuck1010, UtherSRG, Uwagradstudent, Vadim tarakanov, Vald, Van helsing, Vandal B, Vanisheduser12345, Velella, Vicpeters, Vsmith, W1ZzYy, WKLSKE, Warfreak, Wassupwestcoast, Wayward, Weedgarden, Werothegreat, Weyes, Whileships savedhead, Whomp, Wiki alf, Wikky Horse, WillowW, Wilson44691, Wimt, Winhunter, Wknight94, Wlodzimierz, Woer$, Woohookitty, Wshun, Xbows, Xchbla423, XenizE, Xevi, Xexsnowblindxex, Xiahou, Xiong Chiamiov, Y1234567890, Yamamoto Ichiro, Yansa, Zabanio, Zack linick, Zacman78, Zamphuor, Zapvet, Zeman, Zondor, Zotel, 1171 anonymous edits

Biological_classification *Source*: http://en.wikipedia.org/w/index.php?title=Biological_classification *Contributors*: 168..., 172.153.96.xxx, 1984, 2004-12-29T22:45Z, 21655, 5 albert square, Abigail-II, Achowat, Ahoerstemeier, Aksi great, Alansohn, Aldaron, Aleator, Alexei Kouprianov, Alexkin, Aljullu, Allicat07, Allstarecho, Alone Coder, Alpha Quadrant (alt), Ameliorate!, AmiDaniel, Amorymeltzer, Ancheta Wis, Andonic, AndriuZ, AndyCapp, Angr, Animum, Anlace, Anomalocaris, Anomaly2002, Anonymous editor, Antandrus, Antonio Lopez, Apokryltaros, Arendedwinter, Arensb, AshLin, Ashot Gabrielyan, Atif.t2, Aude, AxelBoldt, Bacon and the Sandwich, Baldhur, Ballista, Barkeep, BarretB, Bcasterline, Beano, Because the mail never stops, Beccaa94, Beland, Belg4mit, Bencherlite, Berria, Berton, Betterusername, Billymac00, Biopresto, Blackburn.greg, Blazotron, Blotto adrift, Bob0192, Borislav, Bridesmill, Brockert, BrokenSegue, Brya, Bryan Derksen, BryanG, Bucketsofg, Bugboy52.40, Burner0718, Burntsauce, CJLL Wright, CJTweedy, CWii, Cacao43, Cactus26, Cadaeib, Cadiomals, Calabe1992, Calaschysm, Caltas, Calvin677, Can't sleep, clown will eat me, Canadian Paul, CanisRufus, CapitalR, Capricorn42, CardinalDan, Casliber, Cenarium, Cephal-odd, Chaleyer61, Chilepine, ChongDae, Chorobek, Chun-hian, CiTrusD, Ckampmeier, Clarince63, ClockworkSoul, Closedmouth, Cohee, CohenTheBavarian, Cometstyles, Comm. makatau, Conversion script, Courcelles, Cp420, Cromwellt, Croq, Cureden, Curtis Clark, D, DMacks, DRE, Daft punkette, Dan Koehl, Danarmak, DancingPenguin, Danger, Danielkwalsh, Danilot, Dante Alighieri, David Kernow, David spector, Davodd, Dcattell, DerHexer, Dinoguy2, Dirgela, Discospinster, Disneyfreak96, Dixi, Dlohcierekim, Dmanning, Dobrydneyj, Donald Albury, Donarreiskoffer, Dougweller, Dr.Bastedo, Dragoneye776, DragonflySixtyseven, Dreadstar, Dstar3k, Dwayne, Dyanega, Dysmorodrepanis, E rulez, EWS23, EamonnPKeane, Eastlaw, Edward321, El C, Eleassar, Eliz81, Ellergodt, Emily Jensen, Enviroboy, Eog1916, Epbr123, Eras-mus, Eric Forste, Erkan Yilmaz, Eubulides, Eugene van der Pijll, Everyking, Excirial, Fabartus, Falcon Kirtaran, FeralDruid, Figma, Flewis, Flowerparty, Fluffernutter, Flying Saucer, FreplySpang, Funnyfarmofdoom, Fuzheado, Fæ, GPHemsley, Gaijin42, Gail, Gaius Cornelius, Galoubet, Gandalf1491, Gary King, Gautam p, Gdr, Gearmaster09, Geeoharee, GerardM, Gfoley4, Ghostdood, Giftlite, Gilliam, Glane23, Glenn, Gnostrat, GorillaWarfare, Gracefool, Graham Chapman, Graham87, Granitethighs, Greatal386, Groessler, Grundle2600, Guilee186, Gurch, Gutworth, HJKeats, Hadal, HalJor, Halvard, Hardyplants, Harizotoh9, Hdante, Hectorthebat, Heman, Heron, Hires an editor, Hobartimus, Hodja Nasreddin, Honza Záruba, Hughcharlesparker, Huhsunqu, Hurricane111, Husond, IW.HG, Ian Pitchford, Igiffin, Ilovedulu, IndigoSeptimus, Instinct, Intgr, Into The Fray, Introscop, Invertzoo, Iridescent, Is is Is, Istvan, J. Spencer, J.delanoy, JDspeeder1, Jaknouse, JamesBWatson, Jamesalaska, Jamesofur, Jason.grossman, Jasonandyou, Jauhienij, Jecar, Jerry Zhang, Jh51681, JinJian, Jjmontalbo, Jmeppley, JohnCub, Johnuniq, Joseph Solis in Australia, Josh Grosse, Joymmart, Jpatokal, KNHaw, Karebh, Kartben9, Kathryn NicDhàna, Kbdank71, Keilana, Keith Edkins,

Kel-nage, Kemiv, Kevin.cohen, Khalid Mahmood, KimvdLinde, Kingdon, Kku, KnowledgeOfSelf, Koektrommel, Kotniski, Krawi, Krukouski, Kupirijo, KyraVixen, L11213141516171819, Landon1980, Lanthanum-138, Laookmen, Larryincinci, Lavateraguy, Leandrod, LeaveSleaves, Leeyc0, Legaleagle86, Lenticel, Leptictidium, Lerdthenerd, Lesgles, Levineps, Lexor, LiDaobing, LibLord, Life of Riley, Lindsay658, Little Mountain 5, LittleOldMe, Livajo, Lmc169, Logan, Looxix, Los deits, Lucyin, Luna Santin, M.nelson, MER-C, MKoltnow, MPF, MSJapan, MagneticFlux, Majvr, Mani1, Manuelt15, Manway, Marek69, MarsRover, Master Jay, Mathonius, Maurreen, Mav, McSly, Mdd, Meelar, Mejor Los Indios, Michael Hardy, Mike Rosoft, Minghong, Miquonranger03, MissAlyx, Misza13, MithrandirAgain, Moink, Moomoomoo, Morning277, Mowgli, Mr pand, Mr. Lefty, MrDarwin, Mrhobbz, Mrssandman, Mschel, Muntuwandi, Mwng, Mxn, Myrmecos, N3X15, N5iln, NHRHS2010, Nadiatalent, NawlinWiki, Nayvik, NeilN, NeoJustin, Nguyen Thanh Quang, Nick1nildram, Nicolae Coman, Nihiltres, Nivix, Nneonneo, Norwikian, Notafly, NovaDog, Nsaa, Nufy8, Nunh-huh, Nuno Tavares, ONUnicorn, Obsidian Soul, Ocdncntx, Oemb1905, Ojs, Old Father Time, Old Moonraker, OldakQuill, Oliver Pereira, Omodaka, Onco p53, Optakeover, Orange ginger, Otolemur crassicaudatus, Ouzo, PDH, PM800, PatPeter, Paul August, Peak, Pengo, Perfect Proposal, Peter, Peter coxhead, Petru Dimitriu, Petter Bøckman, Phil1988, Philip Trueman, Phyzome, PierreAbbat, Pihka, Pikachuwashere, Pinethicket, Pippu d'Angelo, Poohbear 0226, Postoak, Prashanthns, Princessamoeba, Proofreader77, Pseudomonas, Pym98, Quantpole, Quiddity, RC-0722, RK, Radagast83, Ranveig, Raz1el, Rds1970, Rdsmith4, Red Winged Duck, Reesyo, Regibox, Remember the dot, Renegadeshark, RexNL, Rgamble, Riana, RichardF, Ridnfrk7, Rizniz, Rkitko, Rocastelo, RogerHyam, Roland2, Romanm, Ronbo76, Ronhjones, Rory096, Rossami, RoyBoy, RyanCross, Salvio giuliano, SchfiftyThree, Scholar1975, Scooter4, Scottalter, Seascapeza, SebastianHelm, Seglea, Semperf, Sentausa, Seqsea, Shafei, Shizhao, Shoes101, Sideways713, Sjwk, Sjö, Smith609, Snek01, Snigbrook, Snowmanradio, Snoyes, Snyper666, Special-T, Specs112, SpeedyGonsales, Stemonitis, Stevertigo, Summer Song, SupernovaExplosion, Syp, THE MIST, TUF-KAT, Taco325i, Tannin, Tau'olunga, Tcatron565, Tdslk, TeaDrinker, Tellyaddict, Temporarily Insane, Tgeairn, The Anome, The Rambling Man, The Thing That Should Not Be, TheBlueFlamingo, Thedjatclubrock, Theresa knott, Tiddly Tom, Tide rolls, Tiggerjay, Tim1357, Timc, Timir2, Tiptoety, Tkinias, Tobby72, Todfox, Tom harrison, Tommy2010, Tompw, Tomtom547, Tony Daly, Tpbradbury, Trusilver, Tuganax, Tvdm, Twinkler4, Ubiq, Ulric1313, Ultimateidiot, Uncle Dick, Uppland, Useight, UtherSRG, Vedantm, Vera.tetrix, Versus22, Violask81976, Virek, Viridian, Viriditas, Vsmith, Wasbeer, Wavelength, Wayward, Welsh, WhiteCat, Wickey-nl, Wik, Wikid77, Wikipelli, Wilke, Will Beback Auto, Willtron, Wimt, Wind, Windchaser, Wisdom89, Wknight94, Wlodzimierz, Wulgulmerang, Xionbox, Yelloeyes, Yeom0609, Yosri, Ytrottier, Zariane, Zidonuke, Zigger, Zoicon5, ZyaX, Zzuuzz, `erin`, Гатерас, ТимофейЛееСуда, Шизомби, 1364 anonymous edits

Flowering_plant *Source*: http://en.wikipedia.org/w/index.php?title=Flowering_plant *Contributors*: (, 2004-12-29T22:45Z, 4twenty42o, 78.26, AUG, Abdul raja, Acdx, Adashiel, Addshore, Addthesmell, Agong1, Alai, Alan Liefting, Alansohn, Alexandra lb, Alexei Kouprianov, Alfirin, Allstarecho, Alnokta, Andreworkney, Andycjp, Anomalocaris, Antandrus, AnthonyBryars, AquamarineOnion, Aranea Mortem, Arcadian, Arch dude, Arkuat, Arthena, AtticusX, Avicennasis, Avono, AxelBoldt, Azcolvin429, Azhyd, Backslash Forwardslash, Bakudai, Baralum, Bballmaster024, Bbplayer4, Bdiscoe, Before My Ken, Ben Hanke, Bender235, Berton, Betterusername, Blockwizzard, Bob the Wikipedian, Bobo192, Brian Crawford, Brianga, Brooke87, BrownHairedGirl, Bruce Marlin, BrunoRreategui, Brya, Bryan Derksen, Bunchofgrapes, Buthwiki, CSWarren, Caltas, Calvin 1998, CambridgeBayWeather, Camw, Can't sleep, clown will eat me, Capricorn42, CarolSpears, Cashew2, Caster23, Cayman20, Cbrodersen, Cflm001, Cheesetuna, Chkenboy, Chrishy, Chrisjj, Chun-hian, Closedmouth, Cnilep, Codiferous, Colby Farrington, Colchicum, Columba livia, Conversion script, Copsandrobbers, Courcelles, Cripdyke, Csparr, Cureden, Curtis Clark, Cxz111, Cyrius, DARTH SIDIOUS 2, Dan Gluck, Dan Koehl, Darth Panda, Dcoetzee, DeadEyeArrow, Deagle AP, Denisarona, Der Golem, DerHexer, Determinate, Dewert, Diannaa, Dinoguy2, Discospinster, Dj Capricorn, Dmacauley, DomenicDenicola, Don't do this HardyPlants, DrFO.Jr.Tn, DrKiernan, DrMicro, Dragonaxl, Drphilharmonic, Drt1245, Dysmorodrepanis, E smith2000, E2m, EDUCA33E, ELefty, EamonnPKeane, Earthdirt, Eassin, Eclecticology, EconoPhysicist, Eichhornia, ElKevbo, Eleassar, EncycloPetey, Epbr123, Eric Forste, Eric Kvaalen, Etxrge, Euryalus, FCSundae, Fabio99999, FetchcommsAWB, Fillard, Fingerz, Fredgoat, Frogger19961, Furrfu, Gabbe, Gail, Gaius Cornelius, Galoubet, Garcsera82, GaryColemanFan, Geezerbill, Geonarva, Giftlite, Ginger1112, Gioto, Glacialfox, Glenn, Globalphilosophy, Gogo Dodo, GolbatTheSexy, GoneAwayNowAndRetired, Graminophile, Grammarmonger, Gscshoyru, Gwandoya, HJ Mitchell, Haabet, Hadrianheugh, Halcon3324, Hardyplants, Haza-w, Hdt83, Hectorthebat, Heegoop, Helikophis, Heman, Hemmer, Henrikhenrik, Hesperian, HighKing, Hinakana, Hive001, Hopefulromntic, Hulu123456789, Hydrogen Iodide, HyperNovaVII, I hate skool-- --, II MusLiM HyBRiD II, Immunize, Iorsh, Iridescent, Island Monkey, Izybella, J. Spencer, J.H.McDonnell, J.delanoy, JNW, Jackhynes, Jager123, Jaknouse, Jeepday, JerrySteal, Jhbdel, Jim1138, Jmgarg1, Jneves121, JoJan, Joe Jarvis, John254, Jonathan Hall, Jonhen8, Jonik, Joseph Solis in Australia, Josh Grosse, Jrtayloriv, Jujutacular, Jusdafax, Just Another Dan, Karl-Henner, Kazvorpal, Killdevil, Killerrabbitsontheloose, KimberleyReid, Kingdon, Kingpin13, Kingturtle, Kitachi, Kitkatcrazy, Kleopatra, KokomoNYC, Komahadara666, Kouka, Krl, Kuru, KyraVixen, L Kensington, Lavateraguy, Leana222222222, LeaveSleaves, LedgendGamer, Lenticel, Li4kata, Liamdaly620, Liene, Lights, Lockesdonkey, Logan, Lophotrochozoa, MPF, Macedoniarulez, Magiccast, Mani1, Manop, Marshman, Matthias M., Mav, McSly, Menchi, Mendaliv, Mephistophelian, Mercury, Mgiganteus1, Michaplot, Mike Dallwitz, Mike Rosoft, Mike and taye, MikeVitale, Mikems, Millahnna, Mitsukai, Mivf, Mmaaddiiee, Mmm, Mollymolly9, Moon&Nature, Morgan Wick, Mowgli, Mozzyepic24, MrDarwin, Mrgreen159, Munita Prasad, NJPharris, NTox, Netoholic, Nickj, Nickname97, Nihil novi, NoahElhardt, Nolaman712, Northamerica1000, Notheruser, Nsaa, NuclearWarfare, Orphan Wiki, PDH, PGWG, Pagingmrherman, PaleCloudedWhite, Pcbene, Pekinensis, Pengo, Penusofnazarethpart2, Peter coxhead, Pethan, Phantomsteve, Pharaoh of the Wizards, PhilKnight, Philip Trueman, Philopp, PierreAbbat, Pietdesomere, Pikachuwashere, Planet plant, PlatypeanArchcow, Polinizador, Pollinator, Polymerbringer, Poor Yorick, Pro bug catcher, Pseudofusulina, Punarbhava, Purplellama31, PuzzletChung, Quandaryus, Qwyrxian, RAW65, Reedy, Renata, RexNL, Richard Keatinge, Richard001, Richardcavell, Rickproser, Ritiksac, Rjwilmsi, Robert Foley, Romanski, Romeslayer (usurped3), Sky Attacker, Slon02, Smith609, Snigbrook, SoCalSuperEagle, Sortior, SpeedyGonsales, Srushe, Stanhopea, Stemonitis, Sterio, Steven Zhang, Suffusion of Yellow, Sugarcoated817, Superm401, Svetovid, TUF-KAT, TaraBartolec, Tau'olunga, Tcncv, Template namespace initialisation script, Terraflorin, The High Fin Sperm Whale, The Thing That Should Not Be, TheodorCH, Thepcnerd, Tide rolls, Tim Ross, Tinymouth, Tom Radulovich, Tom harrison, TomPhil, Tomfy, Tomi, Toroca, Total Tom, Tpbradbury, Transope, Traveler100, Triona, Trusilver, Ufwuct, Uncle Dick, User A1, UtherSRG, Vald, Vanderesch, Vanished user, VashiDonsk, Veddharta, Versageek, Versus22, Vetoeyou, Vlmastra, Voxii, Vrenator, Vsmith, Vuong Ngan Ha, Vzb83, Waggers, Wikimachine, Willking1979, Wimt, Wingchi, Wknight94, Wlodzimierz, Woohookitty, Wotflowers, Writtenonsand, Wtmitchell, X!, XJamRastafire, Xndr, Xskipx7777, YebisYa, Yosri, Yoyospaghettio, Zanimum, Zeamays, Zeman, Zidonuke, Ziegelangerer, , 750 anonymous edits

Image Sources, Licenses and Contributors

GNU Free Documentation License Version 1.2, November 2002 Copyright (C) 2000,2001,2002 Free Software Foundation, Inc. 59 Temple Place, Suite 330, Boston, MA 02111-1307 USA Everyone is permitted to copy and distribute verbatim copies of this license document, but changing it is not allowed.

0. PREAMBLE

The purpose of this License is to make a manual, textbook, or other functional and useful document "free" in the sense of freedom: to assure everyone the effective freedom to copy and redistribute it, with or without modifying it, either commercially or noncommercially. Secondarily, this License preserves for the author and publisher a way to get credit for their work, while not being considered responsible for modifications made by others. This License is a kind of "copyleft", which means that derivative works of the document must themselves be free in the same sense. It complements the GNU General Public License, which is a copyleft license designed for free software. We have designed this License in order to use it for manuals for free software, because free software needs free documentation: a free program should come with manuals providing the same freedoms that the software does. But this License is not limited to software manuals; it can be used for any textual work, regardless of subject matter or whether it is published as a printed book. We recommend this License principally for works whose purpose is instruction or reference.

1. APPLICABILITY AND DEFINITIONS

This License applies to any manual or other work, in any medium, that contains a notice placed by the copyright holder saying it can be distributed under the terms of this License. Such a notice grants a world-wide, royalty-free license, unlimited in duration, to use that work under the conditions stated herein. The "Document", below, refers to any such manual or work. Any member of the public is a licensee, and is addressed as "you". You accept the license if you copy, modify or distribute the work in a way requiring permission under copyright law. A "Modified Version" of the Document means any work containing the Document or a portion of it, either copied verbatim, or with modifications and/or translated into another language. A "Secondary Section" is a named appendix or a front-matter section of the Document that deals exclusively with the relationship of the publishers or authors of the Document to the Document's overall subject (or to related matters) and contains nothing that could fall directly within that overall subject. (Thus, if the Document is in part a textbook of mathematics, a Secondary Section may not explain any mathematics.) The relationship could be a matter of historical connection with the subject or with related matters, or of legal, commercial, philosophical, ethical or political position regarding them. The "Invariant Sections" are certain Secondary Sections whose titles are designated, as being those of Invariant Sections, in the notice that says that the Document is released under this License. If a section does not fit the above definition of Secondary then it is not allowed to be designated as Invariant. The Document may contain zero Invariant Sections. If the Document does not identify any Invariant Sections then there are none. The "Cover Texts" are certain short passages of text that are listed, as Front-Cover Texts or Back-Cover Texts, in the notice that says that the Document is released under this License. A Front-Cover Text may be at most 5 words, and a Back-Cover Text may be at most 25 words. A "Transparent" copy of the Document means a machine-readable copy, represented in a format whose specification is available to the general public, that is suitable for revising the document straightforwardly with generic text editors or (for images composed of pixels) generic paint programs or (for drawings) some widely available drawing editor, and that is suitable for input to text formatters or for automatic translation to a variety of formats suitable for input to text formatters. A copy made in an otherwise Transparent file format whose markup, or absence of markup, has been arranged to thwart or discourage subsequent modification by readers is not Transparent. An image format is not Transparent if used for any substantial amount of text. A copy that is not "Transparent" is called "Opaque". Examples of suitable formats for Transparent copies include plain ASCII without markup, Texinfo input format, LaTeX input format, SGML or XML using a publicly available DTD, and standard-conforming simple HTML, PostScript or PDF designed for human modification. Examples of transparent image formats include PNG, XCF and JPG. Opaque formats include proprietary formats that can be read and edited only by proprietary word processors, SGML or XML for which the DTD and/or processing tools are not generally available, and the machine-generated HTML, PostScript or PDF produced by some word processors for output purposes only. The "Title Page" means, for a printed book, the title page itself, plus such following pages as are needed to hold, legibly, the material this License requires to appear in the title page. For works in formats which do not have any title page as such, "Title Page" means the text near the most prominent appearance of the work's title, preceding the beginning of the body of the text. A section "Entitled XYZ" means a named subunit of the Document whose title either is precisely XYZ or contains XYZ in parentheses following text that translates XYZ in another language. (Here XYZ stands for a specific section name mentioned below, such as "Acknowledgements", "Dedications", "Endorsements", or "History".) To "Preserve the Title" of such a section when you modify the Document means that it remains a section "Entitled XYZ" according to this definition. The Document may include Warranty Disclaimers next to the notice which states that this License applies to the Document. These Warranty Disclaimers are considered to be included by reference in this License, but only as regards disclaiming warranties: any other implication that these Warranty Disclaimers may have is void and has no effect on the meaning of this License.

2. VERBATIM COPYING

You may copy and distribute the Document in any medium, either commercially or noncommercially, provided that this License, the copyright notices, and the license notice saying this License applies to the Document are reproduced in all copies, and that you add no other conditions whatsoever to those of this License. You may not use technical measures to obstruct or control the reading or further copying of the copies you make or distribute. However, you may accept compensation in exchange for copies. If you distribute a large enough number of copies you must also follow the conditions in section 3. You may also lend copies, under the same conditions stated above, and you may publicly display copies.

3. COPYING IN QUANTITY

If you publish printed copies (or copies in media that commonly have printed covers) of the Document, numbering more than 100, and the Document's license notice requires Cover Texts, you must enclose the copies in covers that carry, clearly and legibly, all these Cover Texts: Front-Cover Texts on the front cover, and Back-Cover Texts on the back cover. Both covers must also clearly and legibly identify you as the publisher of these copies. The front cover must present the full title with all words of the title equally prominent and visible. You may add other material on the covers in addition. Copying with changes limited to the covers, as long as they preserve the title of the Document and satisfy these conditions, can be treated as verbatim copying in other respects. If the required texts for either cover are too voluminous to fit legibly, you should put the first ones listed (as many as fit reasonably) on the actual cover, and continue the rest onto adjacent pages. If you publish or distribute Opaque copies of the Document numbering more than 100, you must either include a machine-readable Transparent copy along with each Opaque copy, or state in or with each Opaque copy a computer-network location from which the general network-using public has access to download using public-standard network protocols a complete Transparent copy of the Document, free of added material. If you use the latter option, you must take reasonably prudent steps, when you begin distribution of Opaque copies in quantity, to ensure that this Transparent copy will remain thus accessible at the stated location until at least one year after the last time you distribute an Opaque copy (directly or through your agents or retailers) of that edition to the public. It is requested, but not required, that you contact the authors of the Document well before redistributing any large number of copies, to give them a chance to provide you with an updated version of the Document.

4. MODIFICATIONS

You may copy and distribute a Modified Version of the Document under the conditions of sections 2 and 3 above, provided that you release the Modified Version under precisely this License, with the Modified Version filling the role of the Document, thus licensing distribution and modification of the Modified Version to whoever possesses a copy of it. In addition, you must do these things in the Modified Version: A. Use in the Title Page (and on the covers, if any) a title distinct from that of the Document, and from those of previous versions (which should, if there were any, be listed in the History section of the Document). You may use the same title as a previous version if the original publisher of that version gives permission. B. List on the Title Page, as authors, one or more persons or entities responsible for authorship of the modifications in the Modified Version, together with at least five of the principal authors of the Document (all of its principal authors, if it has fewer than five), unless they release you from this requirement. C. State on the Title page the name of the publisher of the Modified Version, as the publisher. D. Preserve all the copyright notices of the Document. E. Add an appropriate copyright notice for your modifications adjacent to the other copyright notices. F. Include, immediately after the copyright notices, a license notice giving the public permission to use the Modified Version under the terms of this License, in the form shown in the Addendum below. G. Preserve in that license notice the full lists of Invariant Sections and required Cover Texts given in the Document's license notice. H. Include an unaltered copy of this License. I. Preserve the section Entitled "History", Preserve its Title, and add to it an item stating at least the year, new authors, and publisher of the Modified Version as given on the Title Page. If there is no section Entitled "History" in the Document, create one stating the title, year, authors, and publisher of the Document as given on its Title Page, then add an item describing the Modified Version as stated in the previous sentence. J. Preserve the network location, if any, given in the Document for public access to a Transparent copy of the Document, and likewise the network locations given in the Document for previous versions it was based on. These may be placed in the "History" section. You may omit a network location for a work that was published at least four years before the Document itself, or if the original publisher of the version it refers to gives permission. K. For any section Entitled "Acknowledgements" or "Dedications", Preserve the Title of the section, and preserve in the section all the substance and tone of each of the contributor acknowledgements and/or dedications given therein. L. Preserve all the Invariant Sections of the Document, unaltered in their text and in their titles. Section numbers or the equivalent are not considered part of the section titles. M. Delete any section Entitled "Endorsements". Such a section may not be included in the Modified Version. N. Do not retitle any existing section to be Entitled "Endorsements" or to conflict in title with any Invariant Section. O. Preserve any Warranty Disclaimers. If the Modified Version includes new front-matter sections or appendices that qualify as Secondary Sections and contain no material copied from the Document, you may at your option designate some or all of these sections as invariant. To do this, add their titles to the list of Invariant Sections in the Modified Version's license notice. These titles must be distinct from any other section titles. You may add a section Entitled "Endorsements", provided it contains nothing but endorsements of your Modified Version by various parties--for example, statements of peer review or that the text has been approved by an organization as the authoritative definition of a standard. You may add a passage of up to five words as a Front-Cover Text, and a passage of up to 25 words as a Back-Cover Text, to the end of the list of Cover Texts in the Modified Version. Only one passage of Front-Cover Text and one of Back-Cover Text may be added by (or through arrangements made by) any one entity. If the Document already includes a cover text for the same cover, previously added by you or by arrangement made by the same entity you are acting on behalf of, you may not add another; but you may replace the old one, on explicit permission from the previous publisher that added the old one. The author(s) and publisher(s) of the Document do not by this License give permission to use their names for publicity for or to assert or imply endorsement of any Modified Version.

5. COMBINING DOCUMENTS

You may combine the Document with other documents released under this License, under the terms defined in section 4 above for modified versions, provided that you include in the combination all of the Invariant Sections of all of the original documents, unmodified, and list them all as Invariant Sections of your combined work in its license notice, and that you preserve all their Warranty Disclaimers. The combined work need only contain one copy of this License, and multiple identical Invariant Sections may be replaced with a single copy. If there are multiple Invariant Sections with the same name but different contents, make the title of each such section unique by adding at the end of it, in parentheses, the name of the original author or publisher of that section if known, or else a unique number. Make the same adjustment to the section titles in the list of Invariant Sections in the license notice of the combined work. In the combination, you must combine any sections Entitled "History" in the various original documents, forming one section Entitled "History"; likewise combine any sections Entitled "Acknowledgements", and any sections Entitled "Dedications". You must delete all sections Entitled "Endorsements".

6. COLLECTIONS OF DOCUMENTS

You may make a collection consisting of the Document and other documents released under this License, and replace the individual copies of this License in the various documents with a single copy that is included in the collection, provided that you follow the rules of this License for verbatim copying of each of the documents in all other respects. You may extract a single document from such a collection, and distribute it individually under this License, provided you insert a copy of this License into the extracted document, and follow this License in all other respects regarding verbatim copying of that document.

7. AGGREGATION WITH INDEPENDENT WORKS

A compilation of the Document or its derivatives with other separate and independent documents or works, in or on a volume of a storage or distribution medium, is called an "aggregate" if the copyright resulting from the compilation is not used to limit the legal rights of the compilation's users beyond what the individual works permit. When the Document is included in an aggregate, this License does not apply to the other works in the aggregate which are not themselves derivative works of the Document. If the Cover Text requirement of section 3 is applicable to these copies of the Document, then if the Document is less than one half of the entire aggregate, the Document's Cover Texts may be placed on covers that bracket the Document within the aggregate, or the electronic equivalent of covers if the Document is in electronic form. Otherwise they must appear on printed covers that bracket the whole aggregate.

8. TRANSLATION

Translation is considered a kind of modification, so you may distribute translations of the Document under the terms of section 4. Replacing Invariant Sections with translations requires special permission from their copyright holders, but you may include translations of some or all Invariant Sections in addition to the original versions of these Invariant Sections. You may include a translation of this License, and all the license notices in the Document, and any Warranty Disclaimers, provided that you also include the original English version of this License and the original versions of those notices and disclaimers. In case of a disagreement between the translation and the original version of this License or a notice or disclaimer, the original version will prevail. If a section in the Document is Entitled "Acknowledgements", "Dedications", or "History", the requirement (section 4) to Preserve its Title (section 1) will typically require changing the actual title.

9. TERMINATION

You may not copy, modify, sublicense, or distribute the Document except as expressly provided for under this License. Any other attempt to copy, modify, sublicense or distribute the Document is void, and will automatically terminate your rights under this License. However, parties who have received copies, or rights, from you under this License will not have their licenses terminated so long as such parties remain in full compliance.

10. FUTURE REVISIONS OF THIS LICENSE

The Free Software Foundation may publish new, revised versions of the GNU Free Documentation License from time to time. Such new versions will be similar in spirit to the present version, but may differ in detail to address new problems or concerns. See http://www.gnu.org/copyleft/. Each version of the License is given a distinguishing version number. If the Document specifies that a particular numbered version of this License "or any later version" applies to it, you have the option of following the terms and conditions either of that specified version or of any later version that has been published (not as a draft) by the Free Software Foundation. If the Document does not specify a version number of this License, you may choose any version ever published (not as a draft) by the Free Software Foundation. ADDENDUM: How to use this License for your documents To use this License in a document you have written, include a copy of the License in the document and put the following copyright and license notices just after the title page: Copyright (c) YEAR YOUR NAME. Permission is granted to copy, distribute and/or modify this document under the terms of the GNU Free Documentation License, Version 1.2 or any later version published by the Free Software Foundation; with no Invariant Sections, no Front-Cover Texts, and no Back-Cover Texts. A copy of the license is included in the section entitled "GNU Free Documentation License". If you have Invariant Sections, Front-Cover Texts and Back-Cover Texts, replace the "with...Texts." line with this: with the Invariant Sections being LIST THEIR TITLES, with the Front-Cover Texts being LIST, and with the Back-Cover Texts being LIST. If you have Invariant Sections without Cover Texts, or some other combination of the three, merge those two alternatives to suit the situation. If your document contains nontrivial examples of program code, we recommend releasing these examples in parallel under your choice of free software license, such as the GNU General Public License, to permit their use in free software.